HP £2.50

TRAVEL

A 20th Century Experience

TRAVEL

A 20th Century Experience

John Black

The Book Guild Ltd.
Sussex, England

This book is sold subject to the condition that it shall not, by way of trade or otherwise, be lent, re-sold, hired out, photocopied or held in any retrieval system, or otherwise circulated without the publisher's prior consent in any form of binding or cover other than that in which this is published and without a similar condition including this condition being imposed on the subsequent purchaser.

The Book Guild Ltd.
25 High Street,
Lewes, Sussex.

First published 1992
© John Black 1991
Set in Baskerville
Typesetting by Ashford Setting & Design,
Ashford, Middlesex.
Printed in Great Britain by
Antony Rowe Ltd.,
Chippenham, Wiltshire.

British Library Cataloguing in Publication Data
Black, John
  Travel: a 20th century experience
  1. Asia. Description and travel 1945-
  I. Title
  915.0442

ISBN 0 86332 634 X

CONTENTS

List of Illustrations		7
Chapter 1	Early Days and Education in Scotland	9
Chapter 2	China Coast with Butterfield and Swire	26
Chapter 3	Shell Along the Yangtse	41
Chapter 4	Price's Candle Factory and Tientsin	77
Chapter 5	Australia and Return to East Asia	86
Chapter 6	Incarceration During War in the Pacific	91
Chapter 7	Following Japanese Defeat in Asia and the Pacific	118
Chapter 8	Journey to NW China, Tibet and Gobi Desert, followed by Brief Visit to USA and The Netherlands	125
Chapter 9	The End of an Era — Retreat from Shanghai	160
Chapter 10	Hong Kong, Thailand and Countries of SE Asia	166
Chapter 11	Leaving East Asia 'for good'	183

Chapter 12	Jobs and Travel following Retirement from Shell	197
Chapter 13	International court of Justice, The Hague. Case Concerning the Temple of Preah Vihear (Cambodia v. Thailand) June 1960	220
Chapter 14	Search for Early Maps for the Siam Society	226
Chapter 15	High Flight	229
Chapter 16	Single Point Moorings in Far Away Places	231
Chapter 17	Political and Legislation. The Clyde Port Authority (Hunterston Ore Terminal). Private Legislation Procedure. Scotland Act 1936	238
Chapter 18	A Chinese Connection	240
Finale		242
Lecture at Centre for Far Eastern Studies — London University		244

LIST OF ILLUSTRATIONS

Chinese Junk cut in two. Punishment for piracy	34
British gunboat patrolling the Yangtse, prior to World War II	48
Unusual map of Yangtse Gorges and Rapids	66
Shell tanker *Tien Kwang* in a Yangtse Gorge	68
Thumb mark taken by Japanese for identification	93
Prison camp dentist in action	113
Map of Journey to North-West China	128
Rubbing from Thousand Buddha Caves	139
Fort of Hsi Po on the edge of the Gobi Desert	148
Bactrian Camel, the Gobi beast of burden	150
Black and White Temple plan (Khao Phra Vihar) Thailand	176
Sketch of Single Buoy Mooring for Tankers	198
Map of travels in East Asia where SBMs were installed	203
A Trans-Jordan Journey	208
A letter from John Bartholomew, Edinburgh	216
Poem: 'High Flight'	230
Letter from Admiral W. Radford, US Navy, Washington, D.C.	235
From: *Appendix*	
Sketch of The Thousand Buddha Grottoes, on The Old Silk Road near Tun-Huang, Kansa, China, on the edge of the Gobi Desert	258

1

Early Days and Education in Scotland

The only reason I have for recording this autobiography is to tell a story that may interest future generations of my family.

On my paternal side grandfather John Black was born in Linlithgow on 24 September 1840. He came to Dumbarton later and lived at 9 Castle Street from where, at the age of twenty-seven, he married Clementina Lang, aged twenty-eight, whose parents lived at the aforementioned address. Grandfather took a very active part in the affairs of the town. By trade he was a joiner, latterly foreman joiner, in Denny's shipyard and treasurer of Dumbarton Burgh for a number of years. In 1897 he was President of the local Burns' Club and at the beginning of the century a founder member of Dumbarton Building Society, one of the first in Scotland. He was twice married. My father's mother, his first wife, died 13 May 1903.

Father was born at Castle Street, Dumbarton, 5 January 1873, and was married to Elizabeth Stevenson in 1900. I was born on 12 June 1903, my mother's birthday. Father was a marine engineering fitter in Denny's with special knowledge of steam-turbines. At home he was quiet and we seldom heard him express an opinion on any subject. He only took part in family discipline when mother appealed for his aid when she thought chastisement called for his attention. Outside of family and work, father's activity was lawn-bowls, at which he excelled, being a member of the club team. In a non-active way he was a great football fan.

On my maternal side, grandfather John Stevenson was born in 1837 and married 5 December 1862, aged twenty-six, Mary Kirkwood, aged twenty-one. Both were farm servants working in Campsie in Stirlingshire and their marriage took place at

Haggs nearby. They later came to Dumbarton-by-Kilbarchan in Stirlingshire. Grandfather was a master-carter because of his special knowledge of horses and he was employed as such by a haulier in Dumbarton who housed the family at 10 Strathleven Place where mother was born on 12 June 1875.

Grandfather's success in this job enabled him, some years later, to acquire a dairy. There the family had to plan grazing facilities for a sufficient number of cows to deliver milk all over the town. I well recall the house, shop, byre and stable at 80 College Street, which were so well-known to me in my boyhood days. My robust grandmother had four sons and five daughters. I never knew my grandfather, who died on 3 March 1905, but from mother we learned that apart from the Bible his favourite reading was about the Covenanters, big red hankie in hand the while.

Mother was the strong influence in my life then. She read little but J.M. Barrie, one of her favourites, so much so that these lines, 'Never attribute to your opponent a meaner motive than your own', were often repeated by her, with conviction. While not so sure, Rudyard Kipling's 'If' was another favourite, the last verse in particular. We were a religious family — church twice on Sunday and Sunday school. After evening service we sang hymns at home. These are still very precious memories.

Stevenson was mother's own name and like another Stevenson, one who:

> 'Blew on the drowsy coal,
> Held the target higher, chary of praise
> and prodigal of counsel.'

My mother lived to be one hundred years old. For some unknown reason she did not receive The Queen's telegram of congratulations. This was a disappointment to me.

Brother Willie, two years younger, was a schoolmaster in Scotland's south west with a degree from Glasgow University. Sister Elise was born 15 April 1912, the day the giant Cunarder *Titanic* sank in the North Atlantic after striking an ice-berg. She married Peter, a Church of Scotland minister. Despite years apart in East Asia and latterly 'south of the border', we were a close-knit family. My home town of Dumbarton in mid-west Scotland, on the north bank of the River Clyde, was very much part and parcel of my first twenty-two years.

The nineteenth century saw the town firmly established as a shipbuilding and engineering centre. The well-known tea clipper *Cutty Sark* was built there. The shipyard, now no more, has cradled many famous ships that sailed the seven seas. As further evidence of achievement in the world of ship propulsion, the marine engine of the paddle steamer *Leven* has a prominent place in the town centre, designed by Robert Napier the father of marine engineering and a 'Son of the Rock', in 1824.

The most important landmark overlooking the town and shipyard is the castle. Standing as it does in a commanding position on the summit of a 240 foot rock, where the River Leven joins the Clyde, it could well have been conceived as a watchtower guarding the important pass by Loch Lomond into the Highlands. The castle has a longer recorded history as a stronghold than any other place in Britain.

Finally, in the town of my birth, and in common with many other places which have their outstanding old characters, these three cannot be left out: 'Mary the Bowl Wife', 'Johnny Lennox' and 'Tommy the Monkey'. As far as one can remember of early boyhood days, a good number of the same age group lived in the vicinity of what was known as Hartfield. This was a group of three-storey flats at right angles to each other.

Early schooling from the age of five was in the fee-paying academy, then a part of the old buildings at the rear of the Burgh Hall. My outstanding recollection of primary school days, apart from the three Rs, was the way we had to memorize whole chapters of the Bible such as Psalm 139, Isaiah 53, John 14 and Romans 12, all from the Authorized Version of 1611. A repetition of these chapters was called for without mistake when a prominent minister in the town made his annual visit to the school. At the time the words meant little, but I have since come to treasure them.

Secondary education was for me a little over three years. But apparently I was good enough in the primary school to merit an 'A' grade in the higher school. All the subjects were taught by specialist teachers. By popular choice I was elected prefect for two years. This meant being an example to others in the class, which was made up of pupils from all schools in the town and who were getting together for the first time. For me it was quite an experience, although I never won a prize in any subject.

Memories of some teachers are still quite vivid: 'Cha' in maths, 'Janet' or, as we used to say, 'Jinet', in French and 'D.G.' in English.

As boys we had playing and recreational facilities which were the envy of many. Most of these were above-board but there was one in particular beyond the pale. A double-decker tram depot was close by where we lived and the fun we had in this place beggars description. Once the drivers had left their trams in the depot for the night, regulations called for shutting off the power. Some drivers failed in their duty, leaving a heaven-sent opportunity for us boys to try our driving skills. One night we failed to stop the tram and it crashed through a wall and almost plunged into a burn beyond the depot. The discipline meted out following this happening had better be left unrecorded.

The 'rope-work' burn was in the forefront of our escapades. Vaulting and jumping from bank to bank as well as falling in made this tortuous waterway an adventure playground.

The common, a large area for recreation, provided all that was necessary for football enthusiasts. Many later professionals were nurtured here. It was in a corner of the common, near Hartfield, that one of the town's benefactors, Brock by name, set aside sufficient from his shipbuilding fortune to build swimming baths and other facilities. Here much of my leisure time was spent as long as I lived in Dumbarton.

The adjoining country well away from the built-up areas was always an attraction. My favourite place was the environs of Overtown Estate. Access from our home was easy, all the way by footpath, and it was a delight to get there. In spring wild flowers were in abundance and we got to know their names through gathering them and then pressing them in a collection. Daffodils were for mother to use for decorating the home. There was a variety of trees in the glen and the habitat provided for a number of animals: deer, squirrels, stoats, weasels and rabbits. But the Overtown Burn was for me the outstanding feature of the glen. The fall from its source in the Lang Crags, a ridge of hills about two miles distant, made for splendid pools where no swim suit was required. The art of 'guddling' was another attraction which we boys acquired after much practice. Using the hands to catch fish by groping below stones required not only cunning but skill.

As a family we enjoyed summer holidays at a number of

places, but St Andrews was the favourite. There on the Duffer's Course I made my first acquaintance with golf, using only a driver iron and putter. Of course my favourite pastime, swimming in the pool or off shore, claimed most of my time.

The Boy Scouts, founded by Baden-Powell, appealed to me, particularly the Sea Scouts. The troop I joined at twelve had an association with the YMCA and we were fortunate in having good officers and instructors. A yacht, anchored in the River Leven, was a great asset. On weekends we sailed down river to the Firth of Clyde and saw something of the ports at what was known locally as the 'Tail of the Bank'.

In mid teenage years I was patrol leader and finally scoutmaster. The various skills for which badges are awarded brought me to King's Scout level — the peak of Scouting. In 1919, representing town and country, I was the chosen representative for the Scout Jamboree at Olympia, London. We camped at the Old Deer Park, Richmond. Scouts from Scotland also took part in the Highland Games' Exhibition at Olympia. It was an international gathering, each country contributing by presenting a characteristic display. Indeed it was a memorable occasion. It was my first journey out of Scotland and it was by no means the last.

Summer camps with the Scouts were memorable times. The rain and midges of Lochgoilhead, the beauty of Balmaha and Loch Lomond, were entirely different from the east coast. Here Anstruther, in the Kingdom of Fife, then North Berwick, a stone's throw from the Bass Rock, were highlights. But it was Coldingham, in Berwickshire, that came out on top, with its swimming gala in St Abbs harbour and the companionship among the Scouts of the 3rd Dumbarton Troop.

The First World War ended when I was fifteen. It was a trying time for many, especially parents. Boys I knew well, a little older than myself, volunteered to fight for King and Country and did not come back. One I was to hear about many times was my future wife's elder brother, Willie. He joined the Royal Engineers without letting his parents know. After a few months of training he was sent to France as a non-commissioned officer. He was seriously wounded but recovered and was asked if he would train for a commission. He refused, saying, 'I want to get back with the lads in the trenches'. At the Battle of the Somme two years later, while on patrol with three others, all were killed by shell fire. Willie was twenty-three years old.

I was young, and because I had had some responsibility, in the Scouts, two opportunities presented themselves. The most attractive, my first choice, was the coastguard station at Brora, on Scotland's north east coast, where they were in need of watch-keepers for lighthouse duty. This was really a job after my own heart. But it was not to be! My parents decided that I was too young to be so far away from home. The second opportunity was as an orderly in the local hospital, taking care of wounded soldiers. The hospital was to be my contribution to the war effort. I spent many hours and gained lots of experience cleaning up and attending to the minor wants of the 'boys in blue'.

I cannot leave the teenage period of my life without a swimming story. My father was a keen football fan, and played first-class football. I had tried to persuade him to come and see me play in a work's league team. Imagine my surprise when I got home after the match and Father said: 'I saw you. Stick to swimming! You are no good at football.'

It was perhaps due to this parental 'straight-from-the-shoulder advice' that I put a lot into swimming. As a forward in Dumbarton Amateurs' team, I played water polo in the senior Scottish League. It was essential to know life-saving. I spent much time at it, and after obtaining the most advanced certificate in life-saving, qualified as a teacher. There was a time, at an impressionable age, when I seriously considered the professional side of swimming — managing a swimming pool and all that it entails. However, it was not to be. Parental push, and the attractions of a shipbuilding town took me into marine engineering, and that is where this story takes a right-angled turn.

Leaving school at fifteen just prior to what would today be GCSEs, I was employed as an office boy in Denny's engineering drawing office. I did help the family exchequer with the paltry sum of 7/6d a week. Another year at the Academy would have made sense, before qualifying at sixteen, as an engineering apprentice. The year was not totally wasted as I got to know what was involved in drawing office work, though only on the periphery, and something too of what was happening plus the lay-out of Denny's engineering and boiler shops. It was also a time for getting to know something of the outside world. Evening classes were to further my education for the next six years.

It sounds incredible, but in going around the various working areas with drawings and specifications etc., it was at the blacksmith's shop that I encountered a family connection which mother, in her wisdom, decided should not constitute part of a Hartfield Black's life. Mother's brother, Jimmy, and his son of the same name, were blacksmiths. It seems that the devil drink had something to do with mother's attitude. At any rate in my perambulations around the workshops I met both father and son, and thereafter had conversation from time to time with cousin Jimmy. I did not meet Matt, a younger cousin, until some time later, when his father was killed in a tragic accident at work. In later years Matt was in the hierarchy of the Civil Service as a Permanent Under-Secretary and was finally knighted for his services to the country. I recall on several occasions, after retirement from Shell, being invited by Matt to lunch at the Reform Club in Pall Mall.

The period from 1919 to 1923 was a strenuous time in my young life. Apprenticeships in marine engineering involved three years of practical work in the shops and two years of design work in the drawing office as a draughtsman. Starting work at six am was a shock until I grew accustomed to getting out of bed in the middle of the night. I went with Father to the engineering shop. He always bought a morning paper at the railway station en route. But I can remember, quite vividly, that although we walked over the same roads for years to our place of work Father rarely, if ever, spoke a word.

The work was really in the field of mechanical engineering, involving fitting and assembly of engine parts which supplied motive power for the many ships built by Denny, including paddle-engines for vessels plying the rivers Ganges and Irrawady, large reciprocating engines, both steam and diesel, for passenger and cargo ships known worldwide. In the marine engineering field Denny's reputation for turbine propulsion was well-known. Father's special job was on Parsons' action and reaction turbines, which were geared down from the high-speed turbine to suitable revolutions for the propeller. During my three years in the practical get-your-hands-dirty side of engineering I was employed to work with the journeyman, the fully qualified engineer/mechanic. 'Journeyman' is from the French word 'journlies' meaning a fitter or mechanic who works with his hands. It is only used in Scotland. Many of the journeymen had experience of running and maintaining marine engines at sea. It was a very worthwhile experience being with apprentices of my own age and with men who were not only capable but skilled marine engineers.

Apart from the job of learning a trade, this period of apprenticeship involved association with fellow workers both young and old. In effect there was a need for adjustment while not sacrificing my own way of life. With those of my own age I was good enough at football to be in the Engine Works' team, but in some other after work activities I did not take part. With the older men it was a must to work together and get to know, by using head and hands, the skill required to fit and assemble the parts of a marine engine.

Many of the older men were experienced enough not only in their trade but had sailed the seven seas with the engines they had put together. I recall a story by just such an experienced man with whom I was working. He was sailing the world in a tramp ship and while in San Francisco had earned a few days' leave, part of which he felt should be used to paint the town red. Next morning, after a heavy night, he was woken up by the land-lady and through a haze from his bed-side he noticed all the things in his room were topsy-turvy. He was in despair, and thinking to make amends, told her he would pay for all the damage. At this point she motioned him to come to the window and on looking out he saw much of San Francisco razed to the ground. At this he sighed and said: 'Oh but I could not pay for all *this* damage!' He was looking at the devastation caused by the great San Francisco earthquake of 1906.

Ever since starting as an apprentice in 1919 the ambition to be a draughtsman on design work was always with me. After three years in the engineering fitting shop I took steps to fulfil this ambition by taking the examination for entry into the drawing office. Success enabled me to start a new chapter in the discipline of mechanical engineering.

For a moment now I must leave work and give some account of the education that goes hand in hand with practical work. This very essential part of engineering involved evening classes four nights a week from seven to nine thirty. At the beginning of my apprenticeship it will be recalled that work started at six am. In effect, over five years the working time was ten hours each day, plus two and a half hours, four evenings a week, on further education, for six months of the year. The syllabus and examination papers used by Dumbarton Academy, my evening school, were laid down by Glasgow Technical College, now the University of Strathclyde. The results were important in determining one's engineering future.

Much of the work in the drawing office was based on designs that had already proved themselves but with modifications to fall in line with a specification for a particular job. Denny was licensed to build a diesel engine of the Swiss Sulzer type. Staff relations were quite different from those in the fitting shop. Conversation was restricted to the job on the drawing board.

A most interesting part of drawing office work was, for me, the speed trials of ships. Denny's were famous for their high-speed vessels — twenty to thirty knots — as cross channel and naval ships, using turbines as motive power The trials took place in the open water of the Firth of Clyde. As draughtsmen, it was our job to take readings indicating horse power development and revolutions of the propellers in relation to horse power at varying nautical miles. The most exciting work was on deck when, with stop-watch in hand and an eye on the post-indicators ashore, we clocked the speed of the vessel. Among those with an intense interest and standing immediately behind us, were representatives of the owners. As soon as the vessel's speed reached specification there was a relaxation of tension all round, followed by a loud cheer.

The majority of draughtsmen were well on in years, and I recall one in particular who was a specialist in artistic finish, where this was required, on a drawing. He came from Helensburgh, some ten miles west of Dumbarton. His lunch, eaten in the office, was always cold and sparse, and to emphasize his matrimonial relationship, he was wont to tell the same story after lunch, stressing the words he thought appropriate. 'What do you think? I arrived home last night to see a note from my wife on the kitchen table — You will find your dinner in the oven'.

Looking back on two years in the drawing office, from the vista of many years between, my impression is that the atmosphere was very Scottish — conservative and cautious.

And now to the conclusion of my five years of apprenticeship in the mechanical and design section of Denny's Marine

Engineering establishment. My employer said he would retain my services for £2 per week, less insurance stamp contributions. In such circumstances could I do other than look around for ways and means to leave Bonnie Scotland?

As to my other world outside of laying the foundation for making future progress on this competitive planet, another story is perhaps the best way to illustrate the stage I had reached on this journey. Recently, while visiting part of the family in Scotland, a young relative looked me straight in the eye and sad, 'Tell me, how did you meet my grandpa's sister?' And that's just what I have in mind to tell!

The evening school was essential for teenagers, both male and female, who had already made their essay into the working world. The schooling period was a six months' session, terminating in early April with a dance. Then both students and teachers had the opportunity to get together socially, which is always a good thing. I was on the organizing committee for several years. But it was in 1922 that I met someone who was to have a considerable influence in my life for a very long time. Need more be said than that it was made possible through our association in organizing the dance in 1922. Her name was Rena Yuille. I'll tell you more about her later.

Rena was employed in Denny's shipyard and her forte was income tax; it still is. Later she was a teacher of commercial subjects in the evening school and after leaving Denny's she became secretary to the Rector of Dumbarton Academy.

Introductions to families provided no problems. Hours at home at night created a minor problem for me. My parents insisted that Sunday should be observed and that we should all be home by nine o'clock. As to walks be it in hail, rain or snow, Barnhill was our favourite walk, summer and winter. But we needed days off and good weather to top the Long Crags, a ridge of hills from the brow of which the panorama was beyond compare: Loch Lomond and the mountains of the north, the Vale of Leven, the Firth of Clyde and the Cowal Hills. To get further afield we took to bicycles and boating on Loch Lomond.

In the meantime, plans were being made to seek pastures new. There are nearly ten times as many Scots abroad as at home and conditions of work made it necessary for yet another one to be on the move. China was much in my thoughts. From this distance of time, however, I am unable to be specific about

China's call.

Going to sea as a marine engineer appealed to many in a shipbuilding town like Dumbarton. So many were known around who had been to sea and were now working at home. Then there were those who were home on leave from time to time. It was a subject never very far from my mind in the early 1920s. In the family it was a matter that was discussed from time to time. Contributions to the discussion were made occasionally by those who were home from foreign parts. Where was the best place? How to get there? The BI ships sailed from the east coast of Africa to and from the ports of India and as far east as southeast Asia. Paddy Henderson had voyaged from the home ports to Burma or perhaps the China coast. Then it was always possible to take a look at the inland waters on the rivers of the world where many of Denny's paddle-steamers plied their trade. Another possibility, since the First World War, was the oil industry — the Anglo-Persian refineries of Abadan or Burma Oil, the oldest of the petroleum companies. But Father always said, 'Keep away from oil tankers!' And Mother never liked the thought of her eldest son leaving home. She had seen something of the antics of those who returned from far away places.

This was a period in my life when there was a need to crystallize my thoughts. My relationship with Rena was very much in my mind, and the long parting that would be involved, possibly five years. We discussed it often towards the end of 1924, when it seemed certain that 1925 would be a crucial year for us.

We had many day trips together and one especially comes to mind. Loch Lomond was a stone's throw from our home. Known merely as the Loch, its scenic beauty was always an attraction for a day out. This one memorable trip was to Inversnaid on the quiet eastern side of the Loch: sailing from Balloch at the southern end, by the islands, then to Luss, Balmaha and Rowardenan at the foot of Ben Lomond, and so to our destination. The Loch narrows here and Inversnaid is the starting point for a wonderful horse-driven stage-coach trip to Stronachlachor, on Loch Katrine. That was our object until, comfortably seated in the coach we heard the fare announced: 7/6d. It has to be admitted that even for such a splendid ride, we could not take the price. So it was to be a good few years before we sailed Loch Katrine and recalled Walter Scott's *Lady of the Lake*. We had to settle for a day in the environs of Inversnaid and a picnic among the heather.

There was no doubt that I was destined to leave Scotland. Plans were very definitely taking shape, with Hong Kong as the objective. A friend who was a fellow apprentice with relations in Hong Kong, was able to get an appointment, using their influence, with a shipping company there. This involved a connection with their London office and as a result the booking of a passage by the P & O company, with fare paid to Hong Kong.

I was not so fortunate, although through a friend of my father's in Hong Kong I was assured of assistance on arrival. After that it was my responsiblity to find work. The next step was finance, which my parents were unable to provide. The little I had was far from adequate, so I had to borrow from a friendly neighbour, a guaranteed pay-back period.

I recall, as a Christmas present in 1924, getting a number of Harry Emerson Fosdick's books from Rena and making a resolution to read a small part every day. (Fosdick was the minister of a prominent Presbyterian Church in New York.) With Mother and Rena we went to hear Handel's *Messiah* in St Andrew's Hall, Glasgow, on New Year's Day 1925. Much of our time together was spent on favourite hikes within easy reach of home, and from time to time a bicycle ride further afield. I well recall a cycle trip to Balmaha on Loch Lomond side, then taking a rowing boat from there to Inch Cailloch. Rena really disliked being on the water in a 'wee' boat and Loch Lomond was notorious for sudden wind storms which made boating dangerous. However, we were with two close friends and thoroughly enjoyed a quiet picnic on this lovely island. During days of scouting Balmaha was a high spot for camping and the narrows between Balmaha and Inch Cailloch was the stretch of water selected for our regatta.

The weeks and early months of 1925 passed quickly and plans and a booking were made for me to sail from London. A passport is a necessary tool when you leave these shores, together with all that goes with conforming to health requirements through vaccinations, inoculations etc. I have never taken kindly to an inoculation against typhoid, always suffering something of the fever itself after the injection.

The time for parting has come. I don't recall any farewell parties, but I was well aware that my departure was a major event in the family. As for Rena and me, what of the future? We had made up our minds to become engaged after I had cleared the hurdle of Board of Trade certificates — including a certification of marine engineering competence. In the meantime, a regular exchange of letters was to be our only link.

As I look back on twenty-two years of growing up in a religious home, I leave now for pastures new with these words firmly in mind:

'Let the redeemed of the Lord say so.'

The Voyage of the *Katori Maru* from London to Hong Kong
Saturday 23 May 1925 to Friday 26 June 1925.
Sailing distance: 10,254 miles.

The following is taken from my notebook written during the voyage:

May 23

Arrived London and was met by Bert Fenton. Had breakfast at The Strand Corner House. Visited Bert's quarters at Scotland Yard. Train to Albert Docks; boarded *Katori Maru*, sailed at two o'clock. Wrote letters to mother and Rena, posted by pilot at Dover Bank eight pm.

May 25

Still unable to tackle any food. Sea running very high. Finding it difficult to stave off home feeling. 'Let not your heart be troubled neither let it be afraid.' Spanish coast seven pm. Turned in eight pm.

May 28

The blue waters of the Mediterranean and the lovely rocky Spanish coast, perfect day with glorious breeze. Deck golf a very interesting game. Spent most of the afternoon writing letters and pcs.

May 30

Very warm in Marseilles. Stayed on ship. Letters to Rena and Willie. Arrival of Swiss passenger in same cabin. To city to post letters. I don't think much of Marseilles and will be glad when we -push off'. Leave seven thirty pm. City at night a very fine sight.

2,012 miles from London

May 31

Very fine cool day. Going through straits between Corsica and Sardinia. At ten thirty am sighted snow-capped mountains. Been a gloriously fine day. Another letter to Rena and one to Elise. Sunset extraordinary in a clear cloudless sky. Very warm in the cabin, difficult to get to sleep.

June 1

Up early six am. Ship gliding into Naples bay. Vesuvius belching forth smoke. Naples a lovely town. Cabbies very funny and cheaters. Bathe in St Lucia Bay, super. Bought straw-hat. Letter from Rena and sent a few pcs. Had a walk along the streets, bought some fruit.

June 5

Port Said. Early morning rise to see approach of port. Docked right alongside *British Workman*, J Yuille's ship. (Rena's brother). Visited him, cheered wonderfully. Fine ship and engines. Through canal following *British Workman*. Sent off big letter to Rena and home. Fine night on the canal and bitter lakes.

1,600 miles from Marseilles

June 7

Swimming bath open, very cooling. Two dips. Red Sea water very dense and sore on the eyes. Boat drill. Passed P & O liner *Moldavia* homeward bound. Saw some flying fish. Hope to keep Sunday as well as possible. Very warm.

June 9

Competitions: defeated in golf and quoits, passed into next round of peg-board. Still very warm, plenty of dolphins, flying fish and sea birds. Magnificent sunset — two entirely different colours of blue.

June 10

In the Indian Sea now, cool breeze blowing. Southern Cross in the sky last night. Phosphorescence very fine sight. Slept on deck, much cooler.

June 16

Good breeze blowing. Now in final of shuffle board. Finished off letters to mother and Rena, sent photos of sports' programme. Arrived Colombo three o'clock. Ashore to mail and buy some fruit. Spent two hours ashore in the evening, first experience of rickshaw.

June 17

Colombo. Heard of riots in Shanghai. Pcs home. Walk round town. Ship to sail at ten but delayed until two o'clock. Very heavy deluge of rain swept Colombo harbour. *Orsava*, Orient Line, to take mail home. Sky very heavy with occasional rain, much cooler. Shipped a number of Tamil passengers on deck to Singapore.

5,108 miles from Port Said

June 20

Sighted Sumatra eight forty-five am. Island of fine vegetation, lovely mountains, enshrouded in mist. Very dull day. Fancy dress parade — took part of sailor, some very fine costumes worn by crew.

June 21

Dull with plenty of rain. Getting well down the Straits of Malacca. Sighted first Chinese junk. Passed *Dakat Maru* as seen at Bowling five weeks ago. Arrived Singapore ten am, very green place, hundreds of fine islands with magnificent trees. Very hot. Straits of Malacca — harbour entrance very narrow.

June 22

Splendid harbour with many ships. Goodbye to Australian passengers. Harbour about one and a half miles from town. By tram to Singapore to mail letters. Some fine buildings and good streets. Ship for dinner.

June 23

Hong Kong news not of the best — evidence of strikes. Received letter from friends in Kowloon, Hong Kong. Wrote some pcs and another letter to Rena, posted at harbour post office. Ship's sailing postponed until nine am next day.

Colombo to Singapore 1,707 miles

June 24

Sailed at previously advised time. Some new passengers actually joined ship thirty-three miles out from Singapore at twelve noon. Started a descriptive letter to Scoutmaster Hawthorne. Played a few games of Mah Jong. Very interesting. It requires some thought.

June 25

Bright sunshine. Very warm. Played some deck games. Night extraordinarily humid and warm. Saw my first opium smoker. Slept on deck.

June 26

Fine cool day again with fresh breeze. Haircut and shampoo by Japanese barber for 2/-. Ship made good run up to twelve noon. 340 miles.

June 27

Arrived Hong Kong and met by friends.

Singapore to Hong Kong 1,447 miles

2

China Coast with Butterfield and Swire

No one can arrive in this port without being excited by its beautiful setting and the position of the island and its city of Victoria in relation to the mainland. Since that day, I have arrived in Hong Kong many times and have never failed to be impressed. The pictures I have tried to sketch, from a note-book of the time, of the 'Slow Boat to China' would not be complete without a brief reference to the part played by £.s.d.

Fare to London	2 — 0 — 0
London/Hong Kong (*Katori Maru*)	64 — 0 — 0
Expenses on board & at ports	19 — 0 — 0
Expenses in Hong Kong before employment	30 — 0 — 0
First remittance against borrowing made 11 July 1925	20 — 0 — 0
	£135 — 0 — 0

Paid back borrowed amount in seven months at £18 a month.

The home where I spent the first few days in Hong Kong was hospitable but did not bode well for my immediate future. At that time a strike by Chinese crews had immobilized all coastal shipping and since my object was to be associated as a marine engineer with a British shipping company sailing in the waters of East Asia, present prospects seemed far from good.

At that time there were two shipping companies in Hong Kong, either one of which I would have been pleased to join.

Both had century-old connections with the colony and a solid reputation — Jardine Matheson and Butterfield and Swire. An appointment was made with the latter and, contrary to expectations, with so many ships idle due to striking crews, I was appointed right away. The vessel was on the sugar trade with raw material from the Dutch East Indies to a refinery in Hong Kong. The ship, *Taikoo Wan Pi* by name, was alongside the sugar refinery when I joined her. Due to a striking crew, it was a lazy beginning. However, it provided time to acclimatize to an entirely new life and, incidentally, to see around Hong Kong.

My first job was to get to know those who would be my shipmates while afloat. The officers and engineers were all British. I remember getting some very fundamental advice from the Chief Engineer — a fellow Scot from Dundee but of course a much older man. His words of wisdom were these: 'You will be all right if you keep your mind easy and your bowels open.' These were early days and to adjust was not always easy. My life style was quite different. While working, there was a need to get along as a team, but when it came to leisure time my own way of life was important to me.

All ships flying the Red Ensign in these waters are well served with local crews: deck-hands under a bosun, engine room and firemen, for want of a better title, under a No 1, while a chief steward with staff takes care of kitchen and cabin staff. All this makes for comfortable living on board ship and keeps the vessel in good shape above and below decks. For navigation and efficient running of the engines the officers are responsible. The owners handle crew appointments in a very skilful way. The Chinese concerned are from different areas of this very large country where dialects keep them apart. This tactic has obvious advantages for smooth working on board ship.

I was not destined to make even one voyage on the 'sugar boat'. As the strike ended, the powers that be decided on a transfer for me to one of their new ships. The *Anhui* was turbine driven and much more interesting for a junior engineer. This vessel had long interrupted spaces between decks suitable for the accommodation of as many as 2,000 deck passengers. With many other ships, some of them not nearly so well equipped, the *Anhui* was employed to transport Chinese labour from the coastal ports of China to Singapore. The demand for labour was so great in the rubber plantations and tin-mines of the Malay Peninsula where labour was paid at the rate of $1 per day, then worth 2/6d. Then the Chinese population in Malaya soared to fifty per cent of the 5,000,000 total and in Singapore it was eighty per cent.

Now for a description of what I saw, in nearly two years, of happenings on board the good ship *Anhui*. During this time over 50,000 deck passengers were carried cheek by jowl from the ports of Xiamen (Amoy) and Shantou (Swatow). They came on board with families and all kinds of luggage. Livestock was subject to export-duty but this did not include eggs, with the result that thousands of eggs came on board. When the ship was one day at sea the eggs were dropped at just the right height above the deck, and out jumped a chick. Babies were born by the dozen. I saw cases of leprosy for the first time and opium smoking was a revelation to me. But it was the smuggling of opium in its prepared form of heroin that amazed me.

To escape seizure of opium by Customs' Authority the Chinese crew of the *Anhui* planned ways and means beyond belief to get the 'stuff' ashore. On one occasion the 'boy' who looked after my cabin used a drawer there to conceal heroin until the Customs' Officers had left the ship. Somebody 'split' on him so you can imagine my embarrassment. On another occasion the liquid used in the interior cylinders of fire extinguishers was removed and the empty cylinders filled with 'dope'. Following this the useless appliance was once more returned to its emergency position. But for real scheming, which placed the engineer officers on board in an awkward position, the Chinese in the engine-room worked out a plan while the ship was in dock. This involved co-operation with welders and steel-plating workers in the making of a separate compartment within what is known as the coffer-dam on which the main-engine is mounted. In this isolated section in the bowels of the ship much contraband could be concealed. However, it happened from time to time that someone who had inside information was tempted by the reward and gave 'the show' away. This was revealed in front of a judge in the Supreme Court, Singapore, who in summing up the evidence said it was beyond belief that such an elaborate plan to smuggle opium was unknown to the engineers on the *Anhui*. In this connection I was 'within an ace' of appearing before the judge accused of 'traffic in dope'.

Lest it be thought that no good came out of this great traffic in human kind, I have to say that overseas Chinese were well-known then, and still are, for their remittances home to kith and kin. Indeed, it is quite an item in the Chinese economy. The traffic was not all one way to the land of 'milk and honey'. Quite a number, though not nearly so many, of course, were homeward bound, some to see and share their worldly goods with relatives and others, older, to see the Middle Kingdom again near their life's end.

In passing it is perhaps amusing to record two tales. One bright and warm morning in Singapore, I stepped into the saloon for breakfast. A matter of seconds later I made a hasty exit. For reasons of his own the steward thought it was in keeping with this lovely morning in the tropics to introduce something appropriate to the region in the form of a fruit named durian. About the size of a small water-melon, a number of durians were sliced and lay open on plates suitably positioned to catch the eye, and they were good-looking, but the smell permeated the entire saloon. Alas for me and others, no breakfast that morning. If one can get over the odour they say the taste is delicious. The locals love durians. Years later I encountered the fruit again but that is another story.

The second story comes quickly to the point. After the First World War the Japanese were anxious to make drink and food stuffs suitable for foreign taste. Whisky was an early starter and I well remember seeing a bottle on the market which must have been one of the first, labelled 'Queen George', Fine Old Suntory Whisky.

The mail home from me to Rena, father and mother was weekly, as was theirs to me. The borrowed loan to finance my China venture was paid off in seven months. Thereafter, for all my time in the Far East, a regular monthly sum was sent to my parents who may not have needed it but it made life a little easier for them.

The *Anhui* had regular ports of call. My next ship was a 'tramp', the *Hanyang*, older and smaller than the *Anhui*. The engine was of the reciprocating steam type. In comparison with the turbine, the revolutions were small and the work less arduous. The *Hanyang* was a general cargo carrier whose movements were controlled by availability of cargo and draft of vessel in and out of ports. Overall it was a good transfer for me. It opened up many China Coast ports over a distance of 2,500 miles from the South China Sea to the Yellow Sea. Indeed we were engaged on a number of trips carrying rice from Saigon in what was then known as Indo-China.

At this point it is appropriate to say a word about the weather at sea around the long coast-line of East Asia. The typhoon season spans the months from June to October. In the South China Sea where this fierce storm 'kicks off' the period is roughly June/July. Over the remaining months the typhoon moves north until its strength peters out in the Sea of Japan. This very heavy rain and high-wind-monstrosity of a storm takes the form of a huge circle, many miles in diameter. It moves slowly and at the centre it is dead calm, but from there to the periphery the storm effect is very strong, particularly near the centre, with winds of over one hundred knots per hour and rainfall on shore sufficient to cause severe land-slides. I have been in ships involved in a number of typhoons. At one time in Hong Kong harbour the ship was moored to a typhoon buoy but even then, at the height of the storm with everyone standing by, the ship's engines kept the vessel at slow and half speeds heading into the wind. Even with these precautions, when the storm subsided many ships were seen ashore. At sea everything moveable must be 'tied down' and there is special equipment to batten down cargo hatches to prevent possible flooding and capsizing.

The typhoon 'fills up' shortly after it reaches land and expires, so to speak. These violent storms form in the Pacific region of the island of Guam, then move west then through the Bashi Channel between the north island of the Phillipines (Luzon) and Southern Taiwan in the South China Sea. In these same areas, the North-West monsoon blows strongly in the winter and early spring months. The strength of the wind accompanying this storm is enough to slow coastal 'tramps' speed down to such an extent that the same land-mark ashore can be seen twenty-four hours later.

In my time on the China Coast another hazard was encountered, quite different from the storms just described. Most coastal ports had their own fleet of sailing junks, some engaged in fishing, others in general cargo. For long periods piracy was part of their stock-in-trade and while they had a home port the need for their registration was doubtful. In foggy conditions, a common springtime hazard in these waters, it was necessary for powered ships to slow down. In a go-slow manoeuvre of this kind it was not unusual for the vessel to find itself surrounded by junks. In fact I have been in the engine room under these conditions when Chinese voices from the surrounding junks could clearly be heard, as the sound found its way to the engine room down the ventilator. Is it any wonder then that under these circumstances a boarding party from the junks could easily clamber on board? Suitably armed of course, it was easy for them to take what they wanted and even to set fire to the ship. Fortunately I was never involved in receiving the pirates, but one naturally asks, 'Why not arm the ship?' This was done on some ships but it is a 'pretty kettle of fish' if now such a practice is necessary.

To complete the picture of my early days afloat along the coast line of East Asia, something must be said about the ports of call. It has to be borne in mind that the distance involved along this rim of the western pacific is over 3,000 miles or in terms of latitude from the equator to 40° North. However, for the purpose of this life-story I will restrict my ports of call to the coast of China. The ports of Japan and South-East Asia will later be looked at in the second half of my sojourn in China.

The Government of Haikou (Hoi How) on Hainan Island, south of Hong Kong, could only be seen from a faraway anchorage. Families were taken on board here, to make a better living in the countries of south-east Asia where both language and culture were vastly different. Shantou (Swatow) is a busy port only an overnight sail along the coast north of Hong Kong and another source of labour for Singapore and its hinterland. There will be more in explanation of this need shortly. But a story now of Shantou (Swatow) takes us back to the Opium war of one hundred and fifty years ago. It is said of a Jardine-Matheson ship whose captain was a strict Presbyterian Scot that he refused to sail from Hong Kong to Swatow not because his ship was carrying opium but because it was Sunday.

Xiamen (Amoy) port is on the way north from our last port of call and is in the province of Fujian (Fukien) and the main exit port for Chinese artisans on the way south. Part of the port is on an island called Kulangsu. It is beautifully situated, with other smaller islands in a bay.

There has always been a season of the year when Chinese traders and labourers made their way south by sea, but the migrations in the 1920s when I had some association with the mass movement en route to Singapore and Malaysia were exceptionally large. Chinese labour at this time found its fullest scope in the tin mines and rubber plantations of south-east Asia. Chinese artisans were in demand for urban handicrafts. Chinese merchants and shopkeepers arrived to expand the retail trade. Bringing in Chinese for Malay's mines and plantations and opening the doors to Chinese traders, the British created for themselves a problem of the first magnitude. But this narrative is not connected with politics so there I must leave the Chinese migration problem and return to coastal ports.

Fuzhou (Fooehow) is the next port on the journey north. It is the provincial capital of Fujian (Fukien) at the mouth of the River Min and industrially important, renowned for lacquerware but above all associated with tea and the ocean clippers of the past. The *Cutty Sark*, built in Dumbarton in 1869, must have seen a lot of the port of Fuzhou as she loaded tea for the thirsty in those far away islands of north-west Europe.

The Strait of Formosa separates the mainland of China from the Island of Taiwan. This island plays an important part in my story and will come into the picture later, as will Shanghai, the leading port of China at the time and the port of the Yangtse River. These ports were prominent in my life when I returned to China in a different guise.

In the meantime, northward to the port of Zingdao (Tsingtao) in Shandong province, a popular holiday resort and a busy port. There was no lack of alongside accommodation for ships and Zingdao had the attraction of good food and excellent beaches. Much of the export, particularly ground-nuts, was from the hinterland.

Now, sailing north again, a prominent part of the coast line is the Shandong Bandao which juts out into the Yellow Sea. This peninsula makes it necessary for the vessel to make a right angle turn from steering north to a westward course, which involves a very hazardous manoeuvre when a strong north-east monsoon is blowing. In fact, some powered ships who regularly take this course use sails to stabilize the vessel. Wei hai wei, our next port of call, is a sheltered naval anchorage. It was a convenient location for British ships and its use was established by treaty in the days when China was not in a position to object. As a port it was of little or no importance apart from its naval connection.

Yantai (Chefoo) was our next place of call, another pleasant summer holiday resort. The China Inland Mission had a rest camp and school here. It was an anchorage port and all loading and discharging was done using junks. The authorities were very tough in Yantai when any of the junks stepped out of line in disobeying regulations or if piracy was involved. They simply cut the wooden vessel in two by sawing amidships. In my photographic record there is an example of this punishment.

The last stage of our coastal journey is almost due west from Yantai (Chefoo). The ship leaves the Yellow Sea and is soon in the almost land-locked Gulf of Bo Hai (Po Hai). Into this gulf the Yellow River empties its notorious silt content. The ship then is making for Tangqu (Tangku) at the mouth of the Pei Ho, some forty miles of river from the major port of Tianjin (Tientsin). On sighting land one is greeted by the extraordinary sight of many salt-wind mills. As far as the eye can see the countryside is flat and the incoming tide is channelled into narrow canals. Apart from this, a considerable area is given over to many shallow ponds with their own retaining walls. On what is left of dry land a large number of shaky bamboo frames about twelve metres in diameter and four metres high carry a separate arrangement of sails which revolve with the wind. This assortment actuates a small paddle-wheel contraption which moves the sea water to the ponds where it is evaporated, leaving the much needed salt.

The Pei Ho is a narrow and tortuous river and for some forty miles of its length, to Tientsin, a pilot is required. In our coastal journey of some 2,000 miles from Hai-nan Island to Tientsin there is a considerable climate difference. The storm aspect has already been covered in relation to monsoons and typhoons. But when in latitude 20° further north continental weather conditions take over with well below zero temperatures prevailing. In fact, at times during the winter months, the Pei Ho freezes solid and shipping requires the assistance of an ice-breaker.

Tientsin had the special status of a treaty port. In brief, this provided some of the Western powers with concession rights which ended after the Second World War. However, on my first visit there as a transient on board ship I found the British concession a model, well laid out in streets and parks. It is

A Chinese Junk cut in two as a punishment for piracy.

interesting to note that the information just mentioned was given me on good authority, and that apart from names, in nearly thirty years of communist rule nothing in this concession has been altered. But it was to be some years from the present before, with a wife and two daughters, I became resident in Tientsin. And we must wait until then before going into details about North China.

Sufficient now for me to tell a story of a happening in the winter of 1947 on the 'bund' (river front) at Tientsin that will also be vivid in my memory. The river Pei Ho was frozen and ships were ice-bound. For several months, due to crop failure, a vast area of the surrounding countryside was in the grip of famine. It is difficult for any of us to imagine the level of starvation which faced the population there. Families were reduced to eating a soup made from the bark of trees. In such dire circumstances, parents forced young ones of the family to leave home and fend for themselves in the city of Tientsin. In sub-zero conditions, many of these youngsters sought shelter in bales of cotton and other cargo stacked ready for shipment at the port. I well recall witnessing a scene at daylight when with little more remaining in their bodies than a spark of life, many were saved by the Salvation Army arriving with a hot meal and care. From that day onwards I have been a staunch friend of the Salvation Army.

Returning to life afloat in early 1927, I was transferred to another vessel, the *Chengtu*. This ship made it possible for me to extend my coast-of-East-Asia horizon to China's chief coal-exporting town of Qinhuangao (Chinwangtao) close by Shanhaiquan (Shanhaikwan) at the eastern extremity of China's Great Wall where I made a first acquaintance of the monumental structure. In the future, some twenty years later, I was to see the other extremity of the Wall in the foothills of the Tibetan Alps, but that is another story. Then to Manchuria where at the port of Yingkou (Newchwang) we loaded kaoliang, the principal export. The *Chengtu* was a real 'tramp' that enabled me to see the tremendous difference in the ports covering the whole coast line of East Asia, from Manchuria in the north to Singapore and Bangkok in the south, as well as the principal ports of Indo-China, now Vietnam, Saigon (Ho Chi Min City), Tourane (Da Nang) and Haiphong. So it can be said — what a glorious opportunity to see so much in such a short time! While on the *Chengtu* I sat for a Board of Trade Certificate and successfully passed the written and oral examinations. This involved study-time on board ship, but it paid off. Thinking the time was right, I made a proposal of marriage, in writing of course. The proposal being accepted from a distance, it was sealed, if not signed, by an engagement ring.

In mid-1927 a dispute arose among the ships' officers about rates of pay. This led to a shipping hold-up for what seemed at the time an indefinite period. And while I had enjoyed the experience and had been very interested in seeing something of the ports of China's long coast line, this strike provided an important opportunity to move on elsewhere. It was not dissatisfaction with the job but the thought of idleness, even for a short spell, coupled with having no roots ashore, that made me look around for pastures new. An opportunity arose almost at once. The Dollar Shipping Line of San Francisco had a vessel in Japan which they planned to sell in the near future. This complicated shipping deal will be discussed later. In the meantime the *Harold Dollar*, as she was named, needed officers and I was appointed second engineer of this large cargo ship lying in Tokyo Bay. After a pleasant trip from Hong Kong as a passenger on the Dollar Lines *President Jefferson* I caught up with the *Harold Dollar* in Japan. Shortly afterwards this vessel left Yokohama to load a cargo at Otaru in Japan's northern island of Hokkaido for Vancouver. It was something quite different from the China Coast. The Pacific from east to west was our first objective. All hands were new and I had to make friends and get along with all, above and below deck. Some were easy to get on with and others were not so compliant, but it was all in the day's work. As second engineer on a vessel of this kind I was in charge of both engine and boiler rooms. The chief engineer is normally over all but the actual running and maintenance of the power house was my business. At twenty-five years old it was quite a responsibility and that was the way it was on the good ship *Harold Dollar* as we left Japan for the west coast of Canada. In this globe-shaped world of ours the quickest way from point to point is known as the great circle route and we are now set to skirt the Aleutian Islands, much further north than Japan. These islands, part of Alaska, were the only sign of land visible in the voyage east to west.

A pause for reflection now on what might be the reaction back home when the news arrived that I was shaking the dust of the Far East from my feet. Correspondence was regular, but at that stage it was almost two months before one could expect a reply, which meant I was up and away before hearing from Rena or my parents. I am sure the former would be ready to acquiesce, having in mind the prospect of seeing each other much sooner than would have been possible had I stayed in China. Father would have disagreed with my decision to move. For me, in mid-Pacific, I did not know where the *Harold Dollar* was bound after Vancouver but for obvious reasons I had high hopes that it would take me home or near enough to it.

The first time one crosses the longitudinal date-line, 180° east and west of Greenwich, is little more than just noteworthy. It was memorable only on our way from the Orient because we lost a day, unlike the other distinct line on the charge, the equator, a few miles south of Singapore, equidistant from the Poles, where Father Neptune makes sure that his antics beside the ship's swimming pool will not be forgotten.

In mid-ocean we encountered a dead whale encircled by seabirds. I have a photograph of this scene. Every time I look at it I ask myself who passed on the message to the birds indicating the whale's location? The weather generally was good. The chief officer was formerly on the China Coast so we had much in common. Both of us were on the twelve to four watch and at good-weather intervals I spent some time with him on the bridge, usually at sunrise over the wide expanse of ocean. The weather deteriorated as we approached the Canadian Coast and Vancouver Island. Soon it was the Strait of Juan de Fuca, which separates the State of Washington from Canada, then our first port of call in the Western Hemisphere, Nanaimo.

Later, berthed at Vancouver for several days, opportunity was taken to see something of the surroundings which provide such a beautiful framework for this lovely city. Friends who settled here from Dumbarton made my stay the more enjoyable.

The ship discharged all cargo from Japan and all on board were curious to know what was next and for where? Among the officers on board the *Harold Dollar* it was common knowledge that one Robert Dollar, the owner, had recently acquired a large shipping company in the USA. Dollar was a Scot who had made a fortune from lumber on Canada's west coast. In this connection, all his ships were registered in Canada. It seemed to us that in transferring his affairs to San Francisco and now being the owner of a large US shipping company, he would be obliged to sell his Canadian registered vessels. We were not kept in doubt very long.

Our ship was up for sale, but in the meantime, with the pilot on board, we set sail so to speak for the narrow passage between Vancouver Island and the mainland to load lumber at a number of camps. To see the lumber-jacks at work was another world for me. Our cargo of prepared timber for housing was cut to size in the saw-mills. When the cargo holds of the *Harold Dollar* were packed tight with timber in plank-form the ship was far from being on her Plimsoll line. This called for timber to be stacked high above the main deck until the ship settled on her compulsive-load line, clearly defined on the hull by law. The height of timber above the main deck was such that vision ahead from the bridge was just possible and the lumber itself above the deck was chained down to prevent movement.

With this load we set sail for San Francisco, the Panama Canal and the Caribbean. Having no other instructions and with our fuel running low, it was not surprising, when just north of Los Angeles, to receive a signal to 'proceed to San Pedro for bunkers.' San Pedro is the harbour of LA and was close to the oil fields of California. Here fuel was $1 a barrel so at this cost it made sense to 'fill her up'. Unfortunately for me I was in charge of loading and distribution of the oil to a number of double-bottom tanks. This, coupled with instructions to take a maximum quantity, called for my undivided attention which meant that I did not get ashore during my first visit to Los Angeles.

The long coast line of California was almost behind the *Harold Dollar'* when just off the finger of the Peninsula we encountered a force ten storm. This resulted in a somewhat startling happening — the cargo of lumber above deck started to move. As can well be imagined, this caused consternation on board as it resulted in the ship taking a serious list. Fortunately the angle of list did not prevent the ship proceeding after the storm had abated. However, when we limped into the port of Balboa at the Pacific end of the Canal the pilots simply refused to take the Harold Dollar into the restricted waterway of the canal. To get the ship back on an even keel the deck cargo had to be restowed, which took days. The stepping up and down operation in the Panama Canal coupled with the use of mechanical 'mules' for towing makes this waterway perhaps the most interesting in the world. The peak above sea-level is one hundred feet at the fresh water lake of Gatun. We were soon on the way again en route through the Caribbean to Port of Spain in Trinidad, then Bridgetown, Barbados and from there in the Windward group of islands to Guadeloupe in the Leeward Isles. These West Indian Islands were new to me and provided all the fascination that one always associates with superb weather, blue seas and gorgeous beaches. The final port of call for the last of our lumber cargo was Cuba, at a port on the north of the island and to the east whose name I cannot remember. There, after discharge, we loaded sugar for Norfolk, Virginia, in the USA.

Norfolk was a key point in our journey. There we were advised that the ship had been sold to a Glasgow company. This doubtless meant that on our arrival at a European port the new owners would make it known what was required of the officers, particularly their salary. The cargo loaded in Norfolk was coal. In effect this meant, as far as I was concerned, that a change was involved in the raising of steam from oil to solid fuel. Something on this later. The *Harold Dollar*, new name as yet unknown, set out eastward from Virginia into the Atlantic towards Europe. The cargo of coal was sold several times en route before we got the message — first Gibraltar and finally Marseilles. The change over from oil-burning to coal in the boilers was not at all a happy one for the Chinese firemen. For me in the engine-room there was a need to maintain a steam-pressure so that the revolutions of the engine propelled a ship's speed of twelve to fourteen knots depending on weather. When the steam pressure fell, due to stoking failure, I was on the spot, which involved a set to with the firemen. So heated was the contretemps at one stage that I faced a red-hot firing tool drawn from a furnace. Needless to say, on that occasion I beat a hasty retreat although finally we achieved the desired pressure of steam.

Marseilles and my decision was made: to leave the *Harold Dollar* and seek pastures new. So, it was a long train-journey ahead, although it was wonderful looking forward to being home again after two and a half years away from family and my future wife. Rena was on the staff of Dumbarton Academy as secretary to the rector and was a teacher of commercial subjects in the Evening Adult Education Centre. As well as this centre of attraction Father, Mother, brother Willie and sister Elise were pleased to see their more experienced elder brother. It soon became clear though, following the first days of happy welcome, that a certain amount of anxiety existed on the part of my parents about my future. For my part I had no doubt that it would be abroad, ashore not afloat, and, if not urgent, the need to lose no time was very much in mind.

Sometimes an opportunity arises on one's journey along life's way, to reflect calmly, to philosophise if you like, on the recent past and take a visionary view of what lies not too far ahead. The brief spell at home, in late 1927 and early 1928, provided me with just such an opportunity. My life style until I took-off for China two and a half years previously could be said to be cloistered. I was very much under the influence of Christian parents in whose footsteps I followed. Baden-Powell's Scout movement also played a part. Strong drink was *verboten*. Indeed, Mother worried quite a bit that I might be tempted by others who did not regard alcohol in the same light as she did. However, she need not have worried on that score. And as to the 'straight and narrow path', I had no wish to change although it must be appreciated that the environment of sailing-the-China coast was poles apart from Dumbarton and it was necessary to take into account the views of others and entirely different living conditions.

Rena was very much in my thoughts. Although the separation of two and a half years was a long one, the tenor of our regular correspondence was such that there was no doubt about the ultimate aim. Now, home again, this was brought closer. But the immediate need was to seek pastures new and in an occupation that appealed and had a future. Two years in China had left me with a hankering to return, but in what capacity? While there I was impressed by the activities of the oil distributor Shell, known in China then as the Asiatic Petroleum Company. However, in applying for a job I could hardly express a preference to work in a particular country, especially to a company with world-wide operations. Then I felt it was wise to have at least another string to my bow. This turned out to be a position in the Sudan connected with river transport on The Nile. As it turned out, the latter application was answered first, inviting me to London for an interview. En route I decided to make a quick call at Shell. This move was fortunate for me as Shell had just written requesting an interview for of all places, a position in North China.

3

Shell Along the Yangtse

Christmas and New Year of 1927 and 1928 with their accompanying festivities were soon over and my date of departure, 22 February, was very much in mind. A condition laid down by Shell at that time made it obligatory for all staff to be unmarried during the first four year contract. This was most disappointing as Rena and I had decided to be married. The long separation and the fact that Rena had to give up her interesting and lucrative work led to a difference in parental point of view, but we were determined to marry, though my parents advised waiting and Rena's urged us to go ahead with marriage plans.

Be that as it may, the decision on marriage was left, as it should be, to those concerned and our marriage was arranged for 11 February 1928. In accordance with custom then, the banns were duly proclaimed and thrice read to give opportunity of objection in the parish churches of Dumbarton and Knoxland. Separate certificates of this proclamation were signed by the ministers and session clerks concerned. The legal document certifying marriage is an extract from the register book of marriages in the district of Old Kilpatrick in the County of Dumbarton. All of these signed and sealed documents and the family ceremony at Dumbuck Hotel on 11 February 1928 confirmed that we were well and truly married. Our honeymoon was spent in Edinburgh. Had we decided to go to the North Pole it could not have been colder. The long parting of four years, that lay immediately ahead of us, was not very much in mind then. But we had the same Christian background founded on what was right and wrong which was to stand us in good stead over the years.

On 21 February we took the train to London and Rena, my wife of only two weeks, waved farewell to her newly acquired husband on the P & O Liner *Morea* going all the way to Shanghai. The voyage of six weeks was uneventful, although unlike my previous passage to the Far East, this time the ship called at Malta, Aden, Bombay and Penang.

Shanghai was the head-office of the Asiatic Petroleum Company (North China) Limited. Shell used this name for all their selling interests along the Yangtse River and north to Manchuria. Similarly, the Asiatic Petroleum Company (South China) Limited, headquartered in Hong Kong, handled all Shell's business in China south of the area mentioned above. After the long journey from London I was only a matter of hours in Shanghai before I was again on board ship. This time it was a river-boat en route to Hankow (Wuhan), 500 miles west along the Yangtse. My previous China experience provided the best of opportunities to see the coast and its ports. Now for the next four years and beyond my home and work was to be in many ports along China's greatest inland waterway.

This was a first time thrilling steam journey calling at many ports, with a chance to see a close-up of all the action. It was a comfortable trip of about four days with the countryside, port and starboard, revealing how the vast majority of the Chinese made a living. The hard work of farming with primitive tools was their lot, geared to the speed of the water-buffalo. On the river itself there was unbelievable activity: sampans and junks fishing and with cargo. Small steam craft carried many passengers on short journeys. Our ship had its complement too, between the decks. Occasionally huge rafts floated by. The volume of timber on these floating hulks was sufficient to carry a village complete with families. To steer and manage this unwieldy mass of timber was, of course, the work of the elders. It is almost impossible to imagine the problems involved in steering such a vast hulk in this fast flowing river. In itself the river activity was fascinating, and as it is encountered, in the years ahead, it will be very much part of my story.

Hankow (Wuchang) at last — first stop from London and a new life awaits the traveller. In China 'Shell' was known as the Asiatic Petroleum Company whose chief competitor was the American, Standard Oil. Transportation, storage and distribution of oil products was the aim, with kerosene predominating or, to use the title of a best-selling novel of the time — 'Oil for the lamps of China'. To make this product suitable for marketing, all installations were equipped with can-making factories capable of manufacturing at least 10,000 five gallon tins every day. Encountering for the first time such a demand for kerosene, 'The People's' source of primitive lighting, it made me wonder why this vast country was so far behind. More about this as my story unfolds.

But back to the happenings in my everyday life, where four foreigners, or in Chinese parlance 'outsiders', were in charge of the operation which included a labour force of 150 Chinese. It will be readily understood that the need to get on with the Chinese who did the work was important as they were very much part of the team, although separated from us foreigners not only by life-style but language.

Living in Central China in the period between the World Wars was pleasant and agreeable. Servants added to the comfort and the work was not arduous. A tennis-court in the installation provided an opportunity to keep fit. Two of us, with sometimes another, lived in the 'mess' above the office. Our Chinese cook and 'boy' looked after us so well it would not be an exaggeration to say we were spoilt. All expense was shared equally but as far as I was concerned anything stronger than tonic-water was not included. The weather was kindly though hot and humid in July and August. In winter the temperature fell to zero but seldom below.

The storage capacity of our plant at Hankow (Wuchang) was roughly 50,000 tons. But from time to time, as the product range increased, new tanks were erected, which involved, on my part, supervising on-site construction and, on completion, tank calibration. Later, when in use, calculation of oil quantities through measurement temperature and specific gravity. Tankers of 10,000 tons capacity were capable of making the passage of 600 miles, during the high water season, from the Yangtse estuary to Hankow. It was part of the job to see this bulk cargo discharged as it was in the interval between ocean-going vessels to load smaller company-owned craft and railway tank-cars. Distribution of kerosene in bulk and package was over a wide area by river, road and rail. The manufacture of containers in the form of tins or cans of capacity five gallons from flat tin-plate, all the way from Wales, was another responsibility. These containers were used to transport water and other liquids afterwards. In fact, they were very much part and parcel of village life in China where their secondhand value was an important item in the economy.

A dominant feature in our lives was the great river Yangtse. In the height of summer the river rose twenty-five feet above low-water winter-level. This rise was due to melting snows in the Himalayas and swollen tributaries because of the rains. Floods were common over the flat-land areas, from the sea to 1,000 miles west. My own close acquaintance with the great river was a foolhardy swimming adventure. Swimming was very much part of my life from early days, so much so that after a year of observing its movement I planned to take the Yangtse by surprise and swim across. The time chosen to make the crossing was the early summer of 1929 when the river was just on the rise. An essential part of the plan was to finish the crossing at my home base, the oil installation on the north bank. In effect this involved a decision on where to get into the river up-stream on the south bank. To make progress over this straight line distance, from bank to bank, of only one mile, swimming against a down-stream current of five knots, depended on the swimmer's stroke. This was not easy to gauge, which made the decision on the starting point difficult. However, after discussing it with the Chinese crew of our launch, who were very river-conscious, I decided to start the swim on the south bank four miles up-stream from my estimated finishing point. The launch accompanied me. The water was yellow with silt and other foreign matter. The strong current quickly made itself felt and I was being carried down-stream despite swimming with a steady stroke for the opposite bank. There were no incidents of any kind worth recording. But it was apparent after nearly an hour in the water that I would not make dry-land at my calculated home base on the north-bank. In actual fact it was fully one mile down-stream of this point before I touched mud, making the swim across the Yangtse at Hankow (Wuchang) about five miles in fully two hours. I felt none the worse for it though there must have been not a little silt in the alimentary canal.

Thirty years later the Communist Head of State, Mao Tse-Tung, considerably increased his prestige in the eyes of the Chinese nation by swimming across the Yangtse at Hankow

(Wuchang). The time taken is not on record nor how much he had to deviate from a straight line crossing due to current. Suffice it to say that it was a notable event at the time reported in the media world.

Four years was a long spell for a newly married man to be separated from a charming wife. It was made easier by weekly correspondence keeping each other abreast of events. For my wife it must have been tedious though the game of golf helped to pass the time. Not everybody plays golf in the snow with a red ball! For me life was exciting and there was always lots to write about.

When working with Shell abroad it was not part of their obligation to make provision for a holiday of a week or two every year. Depending on location and staff availability, Hankow (Wuhan) was easy enough to reach and considered the place to give staff a break. The mountain resort of Kuling Shan (Jiuling Shan) 100 miles down-river and get-at-able from the river port of Kiukiang (Jiujiang) was an ideal place, 3,500 feet above sea-level. The approach was precipitous and could be made on foot but usually by sedan-chair. I well remember how wonderful it was just to be in the cool a few hours after the humidity of the river valley. It was a joy to walk the paths through the wooded mountain range and from time to time catch a glimpse of the big Lake Poyang. For me this vacation was repeated a number of times, once especially memorably when Fiona our eldest daughter was only six weeks old. The swing of the sedan-chair over the edge of a precipice where the drop was 1,000 feet is still very vivid in the mind of the other person in the chair, holding Fiona: her mother.

After more than a year in Wuhan a German Shepherd dog called Prince came into my possession. He was a beautiful Alsatian and much of my time was spent with him. He travelled with me by river steamer from Wuhan to Changsha — the capital of Hunan Province. The route took us through the shallow waters of the Tungting Lakes where the depth of water allowed ships of only two feet draft to make the passage. The reason for my trip in Changsha was to allow a colleague to go on short leave. It was my first opportunity to be in sole charge of an oil installation and can manufacturing plant — seemed a good omen for my future with Shell, after only two years' service. Prince was great company. We were very much alone but away from the installation it was enjoyable meeting new people and gaining experience in another province.

Hankow (Wuhan) was five miles upstream from the installation and was far enough away by an indifferent road and river to prevent any of us from joining the Country Club which had lots of amenities. The foreign population of this the provincial Capital of Hupeh (Hubei) was considerable. It was a combination of three cities — Hankow, Hanyang and Wuchang where the river Han joined the Yangtse. Later the People's Republic of China joined all three under the name, Wuhan. My connection with this complex was a Sunday Christian service, sometimes at the YMCA and often at the Union Church. In my study of the language, of which more later, the Chinese Church was helpful. I well remember one church where the men sat on one side of the central aisle and the women with children on the other. The preacher was much given to long sermons during which it was quite in order for anyone in the pews to get up, take a walk outside and then come back. Observing a certain restlessness among the congregation the minister invariably said, 'Be patient, wait a little, I will finish in a quarter of an hour.' I well remember these words, even to this day, and how they gave me an insight to the Chinese character.

Another memorable connection I had with Hankow (Wuhan) but in quite a different way was the friendship I built up with a young couple of my own age. The Nelsons were Lutheran missionaries from Minneapolis but their ancestors were close enough to Norway for their roots to be apparent. Dan was great on the tennis-court and we often played at the installation. The friendship lasted for twenty years when it came to a tragic end for both Nelsons, with son and daughter, in an air-piracy incident between Macao and Hong Kong.

Since arrival in China my inadequacy at not being able to speak the language was always with me. Now the opportunity presented itself in Hankow (Wuhan) where I would certainly be staying long enough to get down to study and at least start learning the language. In their wisdom Shell adopted the peculiar attitude of language study in that it was not a necessary tool for engineers. They were prepared to pay only for the tuition of their selling staff. This was quite extraordinary to me who regarded being able to speak the language of the country one lives in as essential. Anyway, it did not deter me so I took the plunge with a teacher from Haungpi — a country town north of Hankow (Wuhan). My teacher did not speak English so there was no escape for me. His long finger-nails stamped him in Chinese eyes as learned, but it was soon apparent that he had never been any further than his native province. In fact, I learned later, when a little more conversant with the language, that my teacher fitted the rather uncomplimentary expression used by the Chinese who were not of Hubei Province: 'In the heavens above there is a nine-headed bird which corresponds on the earth below to the Hubei man.' With Wang 'Hsien Sheng' as my teacher I set out to make progress in mastering this difficult tongue — as I shall relate later. Chinese is a tonal language and as I hoped to be examined on my progress in the not too distant future, my tones should blend with the dialect spoken in Peking (Beijing). The teacher conversed with me in his dialect but I did not know then that because each word has at least four tones or pitch contours my intonation did not meet with the approval of my examiners at the British Consulate in Shanghai. In fact I can still see the faces of the two examiners, wreathed in smiles at my Hubei accent. I passed, so they must have taken my pitch-contours not as seriously as I judged from their looks.

The political scene in China at the time now under review has been left out of the picture so far. However, one could not live along the Valley of the Yangtse between the 'great' wars without being in close contact with foreign-gun-boat diplomacy, and, to some lesser extent, with war lords and their subversive antics to loosen central government control, when they ruled part of the country as regional dictators. The British, American, French and Italian gunboats involved were shallow draft vessels with a four inch gun forward. The ship's company was made up of two officers, part own country and Chinese crew. They were frequently alongside our plant for bunkers so a good rapport existed between us and the officers. On one occasion when a ship of His Majesty's Navy berthed at our installation, it was obvious from the hustle and bustle on board that something beyond the usual was afoot. When I had the opportunity to make an enquiry this is what came to light. The gun-boat had two very important passengers on board, travelling incognito with what appeared to be their ill-gotten gains. They were, to use a term which fitted them very well, 'war-lords', who had surrendered to central government. But to get down to the nitty-gritty, the legitimate government in Nanking (Nanjing) had requested the United Kingdom to use one of their gun-boats to pick up the undesirables from the province astride the River Yangtse, where they had established themselves as regional dictators, and take them to a part of the country where they could make no further trouble. Hard to believe but not uncommon in these parts.

A British gunboat patrolling the Yangtse, Prior to World War II

In the spring of 1930 I was transferred to Chinkiang (Zhenjiang) in charge of the installation there. This port was only fifty miles from the nation's capital Nanking (Nanjing) and 120 miles from the estuary of the Yangtse. Shanghai was get-at-able by train in a few hours. As an oil installation, Chinkiang was smaller than Hankow (Wuhan) but it was in a key position for distribution. This town was on the south bank of the big river and at the junction where the Grand Canal meets the Yangtse.

A word now on the Grand Canal as it is a prominent feature in my life over the next two years. The great waterway starts at Hangchow (Hangzhou) on the gulf of the same name and makes it way for almost 1,000 miles to Peking (Beijing). The Canal was given the ordinary name of Yun Ho — a waterway for transhipment. In actual fact, however, it was one of the most important means of north/south communications in China. It was the Mongol Emperor Kublai Khan who finally linked up the canal as one continuous waterway, some 1,200 years after construction started, in order to guarantee his rice supply. The oil installation at Chinkiang (Zenjiang) was quite different from Hankow (Wuhan). It was part of the town itself and was actually cut in two by a public creek big enough for shallow draft boats. As a fire-hazard it was a 'bomb'. I arrived here with my faithful companion Prince. We were very close as master and dog and he was a great asset in many ways, particularly security. Alas, I was not to have him very long. One day he made an attack on a Chinese carrying a heavy load. I shrugged off the incident thinking it wouldn't happen again, but it did, several times. Finally I got the message that trouble was in the offing so, faced with no other alternative, no vets in this neck of the woods, I had to shoot Prince.

And now to the everyday happenings and outstanding events in my two years' stay here. The foreign business community was small and the club was made to match, with only a billiard table, bridge facilities and a bar. It was not my cup of tea and although membership was a must I seldom visited the club. At one time during my stay in Chinkiang (Zhenjiang) I was attracted by the construction of a major road just outside the city boundary. The road-makers were about to cut their way through a large cemetery. To the Chinese, cemeteries are sacrosanct because they are part of the ceremony associated with ancestral worship. Would the local people be sufficiently worked up to make trouble? In actual fact they were quite calm. It is an old custom in China to ensure that the deceased have sufficient 'worldly goods' with them to assist their passage from this world to the next. At this point I have to admit being drawn to this morbid excavation by a desire to see what worldly goods were being unearthed and whether if artistically attractive, purchase would be possible, as indeed it was.

A resumption of Chinese language study was essential and in this connection a search was made for a teacher whose accent was close to what was known then as Mandarin or Kuo Yo — national speaking. I did not quite get an 'inside-the-wall' Peking

(Beijing) speaker but Miao, my new teacher, was from Shantung (Shandong) province, next door to the old capital. He insisted that early morning was the best time to study, so lessons, three days a week, started just after six am. Progress was good. Miao was much younger than my former teacher and did not subscribe to the traditional Chinese way of life. For him the ancient philosophers, Confucius in particular and one who came later into the picture, Gautama the Buddha, had no place in China of the 1930s. More about him later.

One of the pursuits I much enjoyed in Chinkiang (Zhenjiang), outside of business hours and at weekends, was hunting. The river and the countryside was alive with game. In the former mallard duck in particular was the target and snipe in the marshlands. On dry land golden pheasant and wild boar topped the list. Many hours were spent on the launch and walking the countryside with a shot-gun and good hunting dogs to retrieve the game. Apart from the exercise and excitement the result made a difference to the cook's often lack-lustre menu.

The activities in the installation were many and various. Staff alone was a problem. The watchmen, a very essential part of the staff, came from the local police force, to which they were still attached, making discipline difficult. Fire-drill was important and it was most necessary for all the permanent staff to be involved and to be familiar with the fire-fighting equipment which had always to be at the ready. Five days every week the can-factory produced 25,000 tons for kerosene. These containers were loaded into junks, as manufactured and filled, and were soon on their way across the Yangtse to the Grand Canal and via the lakes in the north of Kiangsu (Jiangsu) province to their destination.

Here, for the first time, I came across steel tanks built into the wooden hulls of junks, ready for carrying kerosene in bulk. Containers in the form of thirty gallon and fifty gallon drums were filled and loaded daily. Shell had a fleet of small tankers operating from Shanghai. These shallow draft vessels, in low water season, supplied bulk to all Yangtse River installations. I was in the fortunate position, with the agreement of Shell, to be employed by the China Maritime Customs' Authority to carry out a survey of passenger launches and tow-boats using Chinkiang (Zenjiang) as their port of registration. This was not only the Customs' people but the Chinese owners. It was a lucrative occupation for me, adding grist to the mill.

A major tragedy for the countryside occurred along the lower Yangtse and its hinterland in the summer of 1931. Due to an exceptionally high rise of the river and unusually heavy rainfall, tremendous floods covered the farmlands and adjoining towns. The oil installation was flooded to such a depth that we were unable to operate. This state of inundation covered an area from the sea to 500 miles inland and was given world-wide publicity in an appeal for aid. Helicopters brought food to people stranded on highlands above the water and, as I well remember, on a thin strip of dry land along the banks of the Grand Canal. It was at the height of the floods I came across a situation almost beyond belief. Among the relief supplies dropped by the helicopters were large rounds of cheese. Despite the fact that people were starving they would not touch cheese. Among the personalities associated with this great relief effort was Charles Lindbergh who, just prior to the Yangtse flood, was the first air pilot to fly the Atlantic single-handed.

The majority of the foreign community in Chinkiang (Zhenjiang) were associated with Christian missionary societies. I had many friends among this group, being a regular attender at Sunday services. The majority were American but the China Inland Mission was very British and fundamental in their Christian teaching. The missionaries of the CIM were drawn from many countries. They had a language school at Yangehow (Yangzhow) which I visited on a trip along the Grand Canal. It will be recalled from Hankow (Wuhan) days I mentioned the Nelsons, of Lutheran persuasion. On my two weeks' holiday period with the Nelsons we arranged a visit to Peking (Beijing) by rail from Nanking (Nanjing) which opened up a new horizon and was most enjoyable. The manager of Shell's small office in Chinkiang (Zhenjiang) once wrote in a staff report about his installation manager — that was me! 'He comes from very dour Scots religious parents.' No comment need be made on this very-much-beside-the-point statement which had no connection with my ability to do the job for Shell and do it well.

Two eminent land-marks in Chinkiang (Zhenjiang) stand out in my memory — Silver Island and Golden Island. The former was situated in the middle of the Yangtse at a sweeping right-angled bed. The latter, some distance up stream, was about one mile from the river, but not so long ago it was in mid-stream.

Both islands are wooded and in Chinese terminology referred to as 'mountains', so called to make them stand out even although only 200 feet above the river and surrounding flat lands. Both are distinguished because they are local centres of Buddhism. Golden Island is topped by a beautiful pagoda. The lower slopes have many temples in garden settings. But without the isolation of an island site it is far from peaceful. On the other hand Silver Island is quiet and the atmosphere of worship created by the monastery and the demeanour of the monks made it a quiet place well worth a visit, so much so that it provided me with my first inclination to look seriously at Buddhism.

My picture would not be complete without making a brief reference to these enchanting islands as seen through the eyes and brush of Edward H. Cree, Surgeon RN in his journal and water-colour pictures — before the days of photography. Dr Cree was on board HMS *Rattlesnake* at the time of the Opium War 1839-42 and made his acquaintance with Silver and Golden Islands when the Royal Navy entered the Yangtse pursuing the interests of imperialism.

Taken from his Journal: Wednesday 20 July 1842
'*Yangtse Kiang River/Chin-kiang-foo*'
'At this point is a beautiful wooded island, covered with pagodas — Silver Island (Seung-shan) — a steep rock with deep passage for ships on each side, fourteen or fifteen fathoms and two picturesque rocks beyond — and beyond again the fairy-like Golden Island (Kinshan) in the centre of the river with its tall pagoda on top and its sides covered with temples, gardens and trees, the walled city of Chin-Kiang on the left and high blue mountains beyond. Further up the river is the entrance of the Grand Canal. The whole makes a beautiful picture — the noble river, beautiful islands and pagodas.'

I cannot bid farewell to Chinkiang (Zhenjiang) without mentioning it was my pleasure to meet the author Pearl Buck of *The Good Earth* fame and many other novels on China, a number of which were filmed and known world-wide. Her parents were American Presbyterian Missionaries in Chinkiang. Her husband then was Professor of Agriculture at Nanking University. An outstanding feature of the meeting and later in her company with Chinese friends was her very fluent knowledge of the Chinese language.

Homeward bound time was approaching and I was very much

looking forward to boarding ship at Shanghai on the way to seeing family, especially my darling wife, Rena. To play the 'game' Shell's way we agreed to a parting of four years. Today folks would say, 'How could you?' Much may be said these days for and against wives being out at work but in the 1920s and 30s for a married woman to be in a job of work outside the home was almost unheard of. Her place was in the home and in Rena's case very much involved bettering her golf handicap. Anyway, shortly this would not be a problem for us and in the future we would be together, at home first of all and then back to China.

The time for mail to come from the UK to China was five to six weeks, an awful long time to wait for news, personal or business. Of course, to cable was always possible but only for very urgent or emergency reasons. So it was quite revolutionary when in the late 1920s mail via Siberia found its way to Shanghai in three weeks. From a personal point of view it was most welcome to get news from home in what was regarded as such a short time. Little did we realize that in the not too distant future travel and mail would get to East Asia in a matter of hours.

Shortly before leaving Shanghai en route home, friends approached me with the request — would I take their son, aged ten, with me to relatives in Scotland? Having agreed to do so I was soon faced with another request to take a second boy of the same age with me, so I fell for two! This involved day and night responsibility over 10,000 miles afloat, from East Asia to Europe. As I stood on the deck of the P & O Liner about to leave from Shanghai beside me were two boys, not quite sure of their guardian, but knowing full well they were stuck with him for six weeks. Just to make sure of our togetherness their parents had arranged for all three of us to be in the same cabin. And then the final straw that nearly broke the camel's back came from the ship's purser who made the situation clear to me: 'You are responsible for these boys.'

During the voyage they did everything but climb up the mast. Nearing Colombo one of the boys broke out in a skin rash. The doctor said, 'We may have to send him ashore to hospital and you must go with him.' Fortunately it turned out to be less serious, although in Colombo I developed an ear ache through contact with a small sea-slug while swimming off Galle Face beach south of Colombo. This was to be really painful for the rest of the journey.

Getting into the Mediterranean I decided we would leave the ship at Marseilles and take the train overland to London. To crown it all, unknown to me, the boys had their last fling in Marseilles. They left the ship after breakfast and were not seen again until the afternoon at two o'clock. The train was due to leave for London at three pm. I was so glad to see them that all I had planned to say and do evaporated as we struggled with last minute preparations to be on time boarding the train which fortunately was on a railway line close to the ship.

London at last! It was a great relief for me to see John's uncle on the platform. But that was not the last of it. Firmly holding my hand, ten year old John looked up at me saying, 'I am not going with him. Does he know my father?' The other boy disappeared from my life at Waterloo Station, London. Nine years later John joined the RAF and shortly afterwards was killed, possibly in the Battle of Britain.

Rena was in a hotel in Southampton Row for our first meeting after our long parting. An essential meeting was arranged with Shell and an appointment made with an ear specialist to get rid of the bug I picked up in Colombo.

In Scotland we stayed at Beechwood for a few weeks. This was a good 'kick-off' arrangement to see and talk to my wife's parents and family and close enough to Hartfield, the Black family domain. My brother was teaching at school and my sister was engaged in secretarial work in our home town. We soon had the opportunity of being on our own through renting a flat in Maryhill, Glasgow, the property of Rena's brother. It was convenient for us and not far from Dumbarton.

We bought a second-hand car and after some driving lessons on my part, getting around was no problem. It was after the wedding at Glasgow University of an old friend from Scout days, then in the Indian Civil Service, that a car mishap revealed a lack of car handling experience on the driver's part. Before going to the wedding reception we were on our way to be photographed complete with wedding garb. En route, on making a right-hand turn off Great Western Road I did not correctly anticipate the speed of the on-coming traffic and ended up just missing a tram-car coming out of the side-road, but fortunately stopped with the car-bumper just touching a plate-glass window. No one was injured and the car was undamaged. It was, however, salutary for my future behind the steering-wheel of a car.

Our first essay away from family was to the Isle of Skye and by train from Glasgow over the lovely West Highland Railway. We made all the bookings for an early morning start and then slept in. Next morning we made no mistake. We loved Skye, the weather was good and the house in Portree that catered for all our needs was excellent. We had no special thing to do so we followed in the footsteps of Bonnie Prince Charlie and Flora Macdonald and of course spent a day at Dunvegan Castle. Such a visit would not have been complete without the company, in book form at any rate, of R.L. Stevenson whose words on this island, which together with Prince Charlie's 'Sing me a song of a lad that is gone ... over the sea to Skye' are famous.

We were back to Glasgow in time for the wedding of my wife's young brother Bert, a school-mate of mine, and then off again shortly afterwards to St Andrews to join my parents. St Andrews has always been a favourite summer holiday resort for the Blacks from early years. I can recall golfing on the 'Duffers' Course with only a driver, an iron and a putter — all the wooden handles. As for swimming, when the tide was out we enjoyed the pools. In fact, before the opening of the Brock Baths in Dumbarton I was swimming at St Andrews.

The remaining time in Scotland was running short. We visited many familiar places in the west of the country and the car was useful for getting us to the right picnic spot. Loch Lomond side and the Trossachs were favourites — tracing footsteps of courtship years. But soon we had to make plans for the long journey by sea to China. This was my third such voyage and I was looking forward to it with pleasure. For Rena, however, it was her first trip off-shore, a major one, and it involved much thought as well as preparation. Not the least of her concerns was parting from parents for the long spell of four years.

On the way, by P & O liner, one of the surprises was the kind or type of fellow passenger one encountered. It could be said of our life in Dumbarton that it was sheltered but brushing shoulders with those on board destined for places and in occupations from the Middle East to the Far East astonished and even shocked us from time to time. Such is life when you leave home, particularly for the first time. An enjoyable feature of the long journey was the variety of games on deck and the weather was beyond reproach. We met friends at Bombay and

Colombo. Our first shock at seeing so many people sleeping in the streets in these places left an impression that stayed with Rena for a long time.

In Shanghai, Shell advised me I would be at their Lower Wharf Installation in charge of the can-making plant. The five gallon container tin-factory here had a record for the highest daily production in China. I would be living in Shanghai and commuting by launch to the installation on the south bank of the river Whangpu — a tributary of the Yangtse. Our introduction to life in this busy city was residence in The Astor House Hotel for a few days, a brief spell in a certain Miss Craig's boarding house and then, of all places in this enchanting land, Macgregor Road, in the district of Yangtsepoo. In our early time there we went for a walk when after a short distance in the fairly crowded streets Rena recorded a famous 'last words' statement: 'I want to go back to our own place now. I have not seen a face like ours for nearly half an hour.'

Shanghai was unique as a city when we lived there in the 1930s. The international concession, a section of the great city itself, was, strange to say, run by a committee of business men who worked there. Their responsibility was public works, police, services, ie fire brigade, public transport, garbage handling etc., but not customs and matters concerning import duty. The concession housed consulates of almost every nation who took care of problems arising from a large foreign population with extra-territorial rights. Adjoining the international area the French had a concession of their own, adding to the peculiarity of the whole set-up. The Chinese city surrounding the concessions made up the conglomeration shown on the map as Shanghai. The authority to which the latter was accountable was the Government of China in Nanking (Nanjing). On the other hand the international settlement in the middle was accountable to no one. Such a state of affairs could not and did not last too long but while it 'performed' a good job was done.

We joined the Union Church and outside of the services Rena was an active member. Most of our social contact was with people we met there. The Minister asked us to contact an English girl who had just married a Chinese — the couple lived near us. We did so and enjoyed many years of friendship. Indeed, over the fifty years we knew them, with many ups and downs in politics within and outside the family, theirs was a happy marriage.

The big event in our lives was the arrival of two daughters in the years 1933 and 1934. It was said when the news was phoned through to me at the oil installation about the arrival of daughter number two, I informed my colleagues it was a son. I still have to be forgiven for this *faux pas*. With the arrival of Fiona Catherine and June Elizabeth Stevenson, both christened in the Union Church, the even tenor of our lives had to be adjusted to taking care of four not two.

Fiona was only three weeks old when we decided on a holiday to the mountain resort of Kuling Shan (Jiuling). Shan means mountain, already referred to in my Central China days. It really was quite an adventure with one so young. Looking back over the years and thinking about what was involved in travel to get to Kuling (Jiuling), the journey must have inspired Fiona and given her the urge to travel which she still has. The first stage of the journey was by river-steamer, some 450 miles from Shanghai to Kiukiang (Jiujiang), on the Yangtse. This was comfortable but it was a shock for mother when the only water obtainable to wash the baby's clothes was a dirty brown colour straight from the river. From Kiukiang (Jiujiang) the next stage was by car, fifty miles through the rice-fields to the foothills of the mountain and then the adventurous ascent.

Two sedan chairs were ready for us. This design of human-carrier has never been surpassed and deserves a few words before we take off. The chair in the centre is fastened to two long bamboo poles giving the carriage an overall length of about eight metres. When lifted, the poles rest on the shoulders of four Chinese porters — two in front of the chair and two at the rear. Bamboo is strong and flexible, thus making it possible for the porters to swing the chair from side to side. Mother and Fiona were in the first chair and Father with luggage in the second. The mountain climb path was narrow and in stretches steps had to be negotiated. Very often as the chair swung from side to side, according to the rhythm determined by the porters, there was nothing between mother with baby and the valley below. This in itself was frightening. But in addition, due to altitude and corresponding fall in temperature, Mother had to search in her hand-luggage for the kind of clothing required, then put it on baby to keep her warm. What a relief it was to get to the hotel and to feel cool after the heat of the Yangtse valley.

The change of air itself was delightful, making even the long and exciting last stage of the journey worthwhile. We met friends and enjoyed the walks along the woodland paths. Kuling (Jiuling) was sufficiently near one of the famous porcelain producing centres of China, Jingdezhen, to ensure that some fine examples of the art found their way to our hotel where they were on sale. It was too early in our China 'life' for us to appreciate the fine points associated with the art of making porcelain but this was a good introduction to the skill of the Chinese in this connection. After two weeks in a complete change of environment and scenery we were relaxed and ready for the pressures of Shanghai. On the way down to the valley below, with its rice-fields, I decided to walk close to the sedan with Mother and our now month old daughter, Fiona.

Some incidents in our lives at this time are worth recalling, such as when we first lived together in Shanghai, in Macgregor Road. One of our three servants did not get on with the others. One day he left a note saying life meant nothing to him any more and we would not see him again as he was off to commit suicide. A fortnight later he returned beaming, looking and acting as if nothing had happened. The cook was a past-master at advising how short he was of certain important commodities in the kitchen. He was always running out of sugar and tea in particular. In actual fact there was no shortage, it was simply that what had been purchased was never used. These goods and others were being returned to and bought by the purchasing source in original packing and our cook collected his 'legitimate cumsha'.

Getting to know the goings-on was all part of experience of dealing with servants. These were the days when we kept beer and spirits only for guests — our consumption was nil. Our house-boy, just mentioned, was bewildered when he saw the only use we had for whisky was to sterilize the knife before cutting a melon. Fish and strawberries don't usually go together but when we look back to the happenings in our early Shanghai days these two items of food, coming as they did as fresh from Japan, were out of this world.

Our neighbour was a friendly person, much older than us, who spent a lot of time on her own — her husband was a seafaring captain. She felt it was helpful to give Rena the benefit of her experience. 'Have a bank account of your own,' she said. 'Quite separate and unknown to your husband. This way you can do many things without going into "committee".' It was a little complicated when she was away from home. Then her servant brought all the family mail to Rena who at her request separated and retained all bank correspondence addressed to the lady herself. Such is life!

Shell were very considerate when it came to arranging for their staff to get a complete break from the job. Kuling (Jiuling) already mentioned was an example. Within seven months of our visit there we had another opportunity, this time of going to Japan: destination Unzen, near Nagasaki on the south island of Kyushu. We were delighted to travel from Shanghai in a boat built in Dumbarton, our home town, in the early 1920s, the *Nagasaki Maru*. We travelled by bus from Nagasaki, some thirty miles to Unzen National Park 1,000 feet above sea level. It was a super change from the delta of the Yangtse. Unzen was well known for its hot springs so we enjoyed the warmth of a plunge in a mixed bathing atmosphere — a natural for the Japanese. The hotel was excellent and the food on the menu very good.

We were looking forward very much to the arrival of the second young Black, then only a few months from seeing the light of day. Rena must have been in good form in more ways than one as we played golf. During our time at golf a Japanese Amah attended to Fiona's wants. Unzen was famous for its golf-course and while we were there, quietness added to the enjoyment of playing in a situation that would now be impossible in golf-playing Japan. The golf episode would not be complete without reference to bird life on the course. Big black crows would pounce down on the white golf balls and pick them up. Many of them must have adorned the nests far out of the golfers' reach. Our only out-of-the-way adventure in Unzen happened while walking along a narrow woodland path when we were confronted by a barrier of large snakes with their heads up and bodies coiled in circles. Not knowing if their venom was poisonous, we decided discretion was the better part of valour and beat a hasty retreat. One morning we were awakened by an unusual sound. Of all countries in the world Japan was the last where we would have expected to hear the bagpipes and into the bargain playing a Scottish Air. Well that's it for Dai Nippon, on this our first visit.

The next major move for me, in my China experience, was

from Shanghai in Kiangsu (Jiangsu) coastal province to Chungking (Chongqing), Szechuan (Sichuan), in the far west, as the crow flies 1,000 miles, but by the Yangtse, the only route open to us, 1,500 miles. With the river at its lowest low level, the latter part of the journey proved to be quite an adventure. Fiona was eighteen months old, ready for her second Yangtse escapade, and June at six months about to embark on her first.

It was towards the end of 1934 that we set out by river steamer from Shanghai. A word on luggage is inevitable at this point and the only word to describe our baggage is 'impedimenta'. Try to imagine what would be required for two grown-ups and two very young daughters. It should be borne in mind that we were taking off on a two-year stint to a place where many items of food we normally consumed were unobtainable. When I think back on this, our first major transfer with the family, the mind boggles at the amount of luggage in bits and pieces that accompanied us as we set out on our 1,500 miles river journey. If ever there was a need to learn a lesson this was the time. In later movements with Shell, who always supplied heavy household requirements, we travelled light with our inventory worked out from experience.

The first stage of our journey was 600 miles on the lower Yangtse to Hankow (Wuhan). Calling at four ports en route made the voyage interesting but it was uneventful. I recall when alongside the pontoon at Hankow Rena had for the first time an opportunity to see a large consignment of block-tea ready for shipment. Some countries, particularly Russia, like to buy their tea in block form. This calls for compressing the leaves and tea-dust sweepings in to the shape of a brick. In the long run it is a cheaper product. And when it comes to a cup of tea in the home a knife cuts from off the block sufficient for immediate use. In China tea is the national drink served at the drop of a hat wherever you go, without sugar and milk of course.

But, to press on, we changed steamers for this the second stage of our journey. The river level was falling, allowing ships drawing no more than five or six feet to make the passage of 400 miles to Ichang (Yichang). The Yangtse skirted the Tung ting hu, a large lake complex now almost dry, leaving only a narrow channel for ships on the way to Changsha. The surrounding countryside on our stretch of the big river from then on was flat ricefield land centred on the only port of call,

Shasli (Shashi).

Information awaiting us at Ichang (Yichang) was not good as far as the next stage of our journey was concerned. We were in for a delay due to low water at vital passages. The water-level of the Yangtse over the rapids within a day's sailing from Ichang (Yichang) was so low even ships with a draft of two feet were unable to make the passage with safety. Fortunately, the Shell Installation could comfortably accommodate the Blacks. We had a good host and enjoyed what turned out to be a three weeks' delay. Ichang (Yichang) was the only place in China proper, ie inside The Great Wall, where the Church of Scotland had a mission station, so we enjoyed meeting the folks and hearing about their work.

While awaiting suitable conditions to embark on the final stage of our journey, I took the opportunity of finding out locally what lay ahead of us on the way to Chungking (Chongqing). It is calculated that if the river were harnessed for power just upstream of Ichang (Yichang), nine times the power of the Niagara could be developed. At low-water Ichang (Yichang) is 128 feet above sea-level while Chungking (Chongqing) is 544 feet. Coupled with this, a glance at the map is enough to show that the area drained by the upper reaches of the Yangtse is enormous. From the melting snows of the Tibetan mountains to the monsoon and local rain-storms, the river and its many tributaries carries a vast volume of water. As it passes through the gorges and over the rapids its hitherto peaceful stretches become infernos of turbulence creating conditions for the navigator that can only be described as nightmarish. These vary very much between high and low waters but whether the ship is set to go against or with the current, the eyes of the master and his pilot must be almost glued to the river immediately ahead. Even if it is necessary to alter the steering by tiller, movement information is conveyed to the man at the wheel by fingers of the hand. Another problem drawn to my attention before sailing was anchorage. There are many reaches in the Yangtse Gorges where it is not possible to anchor and being underway at night is out of the question. This makes it essential, in calculating distance, to get to a safe anchorage before light fails. The prevailing condition for our journey was very low water. At such a perilous state of the river Rena and I with two very young daughters were about to experience what it was like to shoot the rapids of the Upper Yangtse.

At last the green light was hoisted, indicating that conditions were reasonably safe for us to make the passage in our high-powered four-ruddered, shallow-draft steamer to Chungking (Chongqing). The pilots were on board. These former junk-masters, it was comforting to know, had a profound knowledge of the river and its antics. We weighed anchor about noon and the initial sailing was pleasant, with farmlands on either side after we passed the city of Ichang (Yichang). Well ahead of us a range of mountains appeared to be a barrier so, as we approached, it was the more dramatic to encounter a right-angled bend in the river, bringing the ship into the first of the great gorges, Ichone by name. The scenery was magnificent with cliffs on either side rising about 800 feet above low-water level. Being late in the year the mountains reflected the many colours of leaves on the trees, adding to the beauty of the scene. Half way through the thirteen mile gorge the river obliged with another right-angled bend which was easily negotiated. Soon we were leaving the bottle neck for a few miles of open country. But the next danger point was only ten miles away, the Crooked River. Well, is it worthy of the name. Bamboo groves on the foot-hills were a feature on both sides of the river.

In one way we were lucky to be making the passage at this time of the year as the Crooked River is seen at its best when the water is low. All ten miles of it are a mass of boulders and reefs. This confusion presents the pilot with an unenviable steering problem. For us it was an added element of excitement but we were glad to leave this contortion behind.

The next hazard was our first rapid, the Ta Tung T'an. This danger point puts the ship at greatest risk when the river level is low and just such a situation existed as we approached but it was safely negotiated. The mountains closed in five miles beyond the first rapid as our ship confronted a second, the King Ling T'an, where a huge rock divided the river into two channels. The passage of this hazard can only be likened to shooting a torrent. The first safe anchorage for the night was at Miaoho, some thirty-four miles from our starting point. The village was hemmed in by towering cliffs and in the calm that prevailed during this moonlit night the effect on all aboard was eerie, almost uncanny.

On the morning of the second day anchors were up early and

within four miles of Miaoho the nasty little low-level rapid Hsin T'an had to be mastered. Years later I had a terrifying experience on this rapid, which will be part of my story as it unfolds. Twisting and wriggling from bank to bank is about the best way to describe the movements of our ship negotiating the Hsin T'an. In such circumstances I was tempted to ask this question of the pilot: 'What are your feelings when the ship, developing full power, appears to be making no headway?' Now, it is appropriate to pass on an unwritten rule very much in the minds of all upper Yangtse pilots: 'Town-bound traffic has the right of way but all motive-power traffic gives way to junks.' For this reason powered ships always wait below a rapid if junks are seen coming down river.

The Black family were enjoying the voyage and although they saw something of the hazards involved from the main deck I was indebted to the Captain of our ship for the details and place names aforementioned and for those still to come.

A rather uninteresting passage of about ten miles lay ahead of us before reaching the picturesque walled city of Kueichou. Five miles away from here the ship was prepared for a tough encounter with a number of rapids. The Yeh T'an was the terror of them all. A large shingle bank on the port side narrowed the channel ahead and one could see quite clearly on approach a build-up of water at the head of the rapid. A call from the bridge put the engines on power to enable the ship to shoot the rapid. Further rapids of little consequence lay ahead and from there for the next six miles the river opened out, with farming scenes on the banks until the cliffs closed in again.

It was at this point that our ship entered the second of the great Yangtse bottle-necks, the Wushan Gorge, between two sheer cliffs. The timing of our voyage in late November, coinciding as it did with very low water, provided in itself a view of the gorge entry seldom seen. As we made our way from broad daylight into shadow, a shaft of light from the mountain tops streamed into the depths of the gorge. To say the least, it was awe inspiring. Some eight miles from the entrance, the narrow gorge opened up a little and cave dwellings could be seen hewn out of the cliffs high above us. We were thrilled with the scenery, which included a small village perched precariously on the cliff-face. Thanks to the beauty of nature, many miles of lovely scenery lay ahead before the gorge opened out for the ship's overnight anchorage at the walled city of Wushanhsien.

The ship made an early start on the third day and for the first eight miles it was plain-sailing. At mileage just short of one hundred from Ichang (Yichang), another big shingle and boulder mound off the river bank forced a nasty build-up of water to form a rapid, so much so that it blocked the view up-river from our ship's bridge and the pilot had to rely on a signal station for a safe run through. For five miles, thereafter, the countryside was open and the going easy. But we were soon to see towering mountains flanking the last and greatest of them all, the Wind-Box Gorge.

Approaching this hazard it was impossible to see any entrance in the 3,000 feet façade which faced the ship. Then suddenly, a sharp corner and we were in the gorge itself. It is said, and the Blacks agree, at least Father and Mother, that no words could describe the magnificence of the next four and a half miles. It was as if 3,000 feet of mountain was split down the middle to form a giant corridor, narrow in places to little more than 100 yards and with walls appearing to overhang the river itself. Recalling this part of our journey, my wife expressed herself in these words: 'As I viewed the awesome grandeur of this work of nature it made me feel humble and lowly.'

We are now clear of the great gorges and for the next thirty miles farm houses, their cultivated lands, green trees and red sandhills take up our attention, an occasional temple situated near river-danger places built by or for junk-masters for their worship and to set minds at rest before embarking on a dangerous stretch of river. For this purpose it should be said that temples and images of the Buddha are common in places of danger, and there are many on the Upper Yangtse. Just 200 miles from Ichang (Yichang) our vessel reached Wanhsien and a firm anchorage for the night. Its situation can only be described as unique, divided in two by a mountain torrent it appears to cling almost insecurely to the side of a 500 feet cliff. Its geographical position made it sensitive to politics, particularly in war-lord times. Wanhsien is famous for two vastly different things: one artistic, filigree silver work, and the other vastly different, wood oil, the base of the best varnish, obtained by crushing ripened nuts from a tree common in Eastern Szechuen (Sichuan).

The fourth day out from Ichang (Yichang) saw our ship 175

miles from Chungking (Chongqing). There were no real thrills in this section apart from a few rapids which were in need of pilot expertise. It was open countryside beyond the banks most of the way and a number of towns enlivened the passing with a multitude of small craft, sampans, in the vicinity.

And now this 2,500 mile journey from Shanghai, which has taken over a month, is almost over. With my wife and two very young daughters we are nearly there at the entrance to the Brass Gong Gorge, small in comparison with the Wind-Box Gorge but our very own which we will be able to see at any time. Our neighbouring village, Tangchiato, is within sight and the oil tanks of the installation are in view. But above all, 300 feet from the river level, is our home for the next two years and from which we have a superb view of the gorge.

Unfortunately, Fiona was quite sick on arrival. Something she had eaten on board ship made it necessary for us to go to Chungking (Chongqing) with her to see a doctor. This involved not only being unable to start work but lengthy journey by launch, twenty miles in all. However, the sickness turned out to be less serious than anticipated and was soon rectified.

Before introducing work in this out-of-the-way place and the assistance I had in getting it done, a word on our house and servants would be appropriate. The house was situated well above the river, indeed on a high place away from installation activities. Shell had designed the residence to compensate for its well-away-from-everything location and to some extent to take care of loneliness. For us the latter was not a problem. Downstairs there was a dining room in close proximity to the kitchen. The sitting room, adjoining, was large and comfortable, having a panoramic outlook to be described later. On the same floor was a bedroom with bathroom, while above, by way of a beautiful stair-way, we had three bedrooms all with separate bathrooms. Verandas provided shade for the rooms mentioned and were equipped with cane furniture for relaxing outside. Our three servants, cook, amah and houseboy, had their various duties and despite the fact that they could not speak a word of English they did a good job for us. The kitchen was large, with a concrete floor which, with our standard of cleanliness, had to be scrubbed in every corner once a week. Apart from his obvious duties cooking for four the cook, made all the bread, using potato water as yeast. The amah had two children to

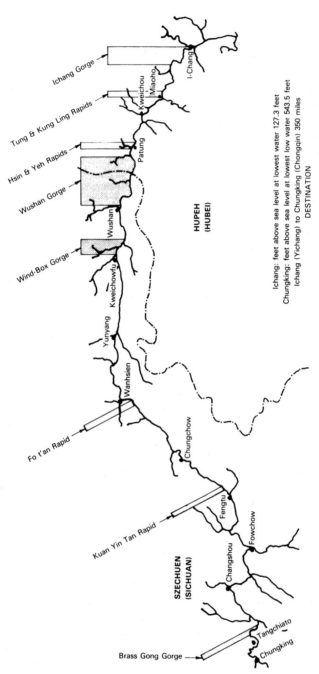

An unusual map of the Yangtse Gorges and Rapids

look after as well as to initiate gossip. The boy had to do everything concerned with keeping the house neat and tidy. In connection with the overall running of the mansion and the servants, the compliment must go to my wife. Still in her infancy of living in the orient and with servants unable to speak her language, she did a superb job in this far-away place.

The installation with its oil storage tanks for various products, kerosene again predominating, making a five gallon fabricating tin-plant essential, employed the majority of workers. But the heart of all was a diesel-driven power-plant for the factory machines, pumps, electric power for lighting and an ice-making machine. The tradesmen were engine-fitters, carpenters, blacksmiths and clerks. These key men and a Chinese, working closely with me, were all accommodated in a special building within the installation. Other employees, and there were quite a number involved in the movement and checking of oil products in bulk and package, lived in the adjoining village of Tangchiato, some ten miles by river from Chungking (Chongqing).

All product movement in and out of the installation was by river. The Yangtse makes a right-angle bend here and is restricted in its flow as it enters the Brass Gong Gorge. Coupled with a tremendous increase in the volume of water in the river due to the melting snows, a source in the Himalayas and rainfall in the area of its many tributaries, the river can rise thirty feet in twenty-four hours. And over the season — late spring and summer — the difference between low and high level is over 100 feet. This changes the contour of the countryside and had to be taken into account when planning the lay-out and operation of the installation. The tanks and buildings were roughly 125 to 250 feet above low water. During the high rise and fall season, just mentioned, the pontoon mooring berth for all craft was under watch twenty-four hours a day and a team was always ready to handle its anchor mooring chains and flexible hose connections. During the season of quick 'freshets', the name given to sudden river rises, all installation staff were on stand-by. It is no exaggeration to say we were in the hands of the great river — nature's dictator! Fully 100 feet by 500 feet of steep sloping river bank, exposed at low level and fronting the installation, was so fertile due to a deposit of rich silt that when the river was high local people were able, in a six months' period, to plant and harvest three widely different types of crop.

Shell tanker *Tien Kwang* in a Yangtse gorge

The transportation of supplies from Ichang (Yichang) was a problem. Oil in bulk was of prime importance along with other problems in drums as well as tin-plate for can-making, to mention but the essentials. Then we needed items of food unobtainable in these parts, for two young daughters and their parents. Shell designed, I am sure, with the assistance of experienced naval architects, three vessels suitably powered with steering capacity and acceptable draft for the gorges. When loaded to a draft of five feet the largest was able to carry 500 tons of oil in bulk and about 100 tons of packed cargo. These ships were suitable for the upper Yangtse and plied between Ichang (Yichang) and Chungking (Chongqing). This was our connection with the outside world, depending on the state of the river. At very low water none of these three vessels was able to negotiate the hazards already described. However, I had known them arrive at the installation with no paint on their funnel. It was all burned off due to an order from the bridge for full power at the highest permissible boiler pressure to enable the ship to get through a series of rapids. In very low water, when we were in need of urgent packed supplies, much smaller cargo vessels were used, similar to the type which brought us here. This was our only way of getting fresh fish, often thrown over the side as the ship passed Tangchiato.

Visitors were so very infrequent that one day when my wife saw two ladies walking up the steps just below the house her imagination took over and she immediately thought they must be from another planet, hopefully heaven. The ladies were missionaries and had walked from Chungking (Chongqing), using byways as highways were non-existent in their long walk. Our only means of communication with the big city, all of ten miles away, was by launch, and when a 'freshet' bedevilled the river the launch was unable to stem the current and we were quite cut off. It was a real joy when our ships berthed and we were able to entertain the officers. Sometimes the entertainment shook our quieter way of living. There was an occasion when one of the lads opened the top of my wife's precious piano and threatened the interior with the contents of a beer mug. There was a pregnant silence and then my wife exploded. However, peace was soon restored and the piano was undisturbed. On one occasion we did have a VIP visit us all the way from Shanghai — Shell's Chief Engineer who appeared to be impressed, particularly with the high quality of our lunch. Perhaps this is not the time to say so but it was not until then, at the age of thirty-two, that I fell by the wayside and sampled beer. In the two years we were at the installation only once did we see the local Shell manager, who lived but ten miles distant. Other members of the staff had to see me on business and we actually had a young missionary couple, as our guests, on their honeymoon. But the prize of all was a visit from the British Consul's wife and the never to be forgotten incident at lunch. Our very alert Chinese house boy, doing his waiting at table, observed that the lady was left-handed and quickly rushed out from behind the screen and changed all her table utensils. Everyone at the table was conscious of the move and she was furious. Indeed, afterwards she suggested to my wife we should sack him. Needless to say, we never did. In fact, not long afterwards we were invited to his wedding in the near-by village, a ceremony in itself not to be missed.

In a journey by launch from Tangchiato to Chungking (Chongqing) the river forms a curve, for all the world like a huge horseshoe, in a ten mile circuit. On the last mile of the journey the waterway was alive with traffic, sail and hand-propelled. It was the big junks on which attention was focused; they were making for, and many had already anchored at, a point where landing was safe and easy. The attraction here was a colossal figure of the Buddha in traditional crossed-legged position. The junk-masters or laodahs with joss sticks in hand, were on their way to say 'thank you' to the Buddha for their safe journey over the rapids and through the gorges. The position of the Buddha, here on a solid foundation and well taken care off, provided all users of the great river with vital information. This man of meditation was seated in such a position that as parts of his body were covered by the rising river so a message was conveyed to all afloat to take care. And when the river water covered his eyes it was no time to be afloat.

My visits to Chungking (Chongqing) were few and far between and then only to the township on the opposite bank of the river, Lungmenhao, where Shell had an office. Above the river level here the hills were wooded and in places cultivated. It was a pleasure to live in Lungmenhao, away from the rough and tumble on the other side. This extraordinary city, the commercial capital of West China, was situated on a steep sandstone reef or peninsula between the Yangtse and the Chialing, one of its principal tributaries. The rise of the river at Chungking (Chongqing) created a major problem. From its low winter to high level in summer the difference was 150 feet. The resulting difficulties were tremendous, not only for river traffic but for the fraternity who simply had to live close by their livelihood. Big junks crowded the river's edge and temporary townships built on stilts at low water provided shelter and a home for what we in the west call stevedores and their families. The same thing happened year after year when the river started to rise. Many of these shacks were swept away in the first flash-flow of the Yangtse, although a tremendous effort was made by the owners to carry the dismantled hovels on their backs to safety before the river made this back-breaking task impossible. When the stilts or piles were long and strong enough, survival was possible, but it was an annual touch and go tussle. The waterfront was always a peculiar sight and as if those who lived there did not have enough to contend with, the rats on that level were as big as cats. Away above this bizarre scene the walled city stood aloof as if it had little or no connection with the roughhouse below. The ascent and descent to and from the city gates was always crowded. Many were the steps in this twenty feet wide steep stairway usually running with water but designed to make it easier for the many engaged carrying heavy loads on their shoulders. I was mixed up in this hubbub but seldom, although from time to time I had occasion to visit friends in the city, using these ungainly steps.

For nearly 100 years, until the Japanese attack on Pearl Harbour, the Royal Navy maintained a presence on the Yangtse in the shape of a flotilla of small gunboats. In point of fact, the presence extended over 1,500 navigable miles of the mighty river. My connection with this minor arm of the Royal Navy was quite considerable as Shell Oil played a definite part in their mobility. Many a story has been told about the escapades of these gunboats and the tale I am about to tell can be described, it seems to me, as no more than a diplomatic prank. Suffice it to say that I vividly recall a bunkering operation involving HMS *Cockchafer* at Hankow (Wuhan) when a very excited ship's company brought their vessel alongside our installation. Acting on behalf of the Chinese Government of the day in an area between Ichang (Yichang) and Hankow (Wuhan), the captain of the *Cockchafer* stopped to take on board two warlords complete with baggage — and there was plenty including the monetary contents of banks in a wide area. The gunboat captain's job was to see these fellows clear of the Yangtse Estuary onto another ship en route south. This tricky affair must have been planned by none other than the British Ambassador in Nanking (Nanjing) at the request of the Chinese Government there. Such was the state of affairs in China in the 1920s.

A second incident, in which the White Ensign took part, was more humanitarian as it concerned all four of us — my wife, self and two daughters, while we lived at isolated Tangchiato near Chungking (Chongqing). I was never altogether happy with the security position there. We were far enough away from Central Government to be outside their authority and while warlords existed in Szechuan (Sichuan), there were pockets where others were a law unto themselves. To some extent this was part and parcel of the goings-on in the neighbourhood where we lived, largely due to cultivation of the poppy, grown to supply the worst of all drugs, opium. Unfortunately, we were close to a down-river transportation point, which in itself led to unlawful practice. One night we were all well and truly sound asleep when a loud knocking at our door brought me in seconds to the source of the noise, to be faced with an anxious watchman who informed me that round our boundary we were encircled by *t'ou fei*, robbers. There was little open for me to do in such a tight corner; family safety was paramount and the property of Shell was in my care. Fortunately, some thought had already been given to the possibility of just such a happening and I was in possession of a Very pistol and cartridges to be used in emergency. The time was now and the Very light equipment was close at hand and the house in an elevated location, perfect for firing. Using cartridges in the colour sequence arranged, I fired hopefully. The watch on the gunboat at Chungking (Chongqing) observed the signal and the bandits too saw the firing of the Very lights and did not press their advantage. Some hours later, much to our relief, need it be said?, HMS *Gannet* dropped anchor off Tangchiato.

Apart from Hankow (Wuhan), where there was a large foreign community, in other treaty ports along the Yangtse expatriates were few in number. One of the joys of having a gunboat at one of these ports was the welcome sight, usually on a Saturday morning, of the Curry Pendant flying from the main mast. This was an invitation for all ashore to join the navy for curry lunch.

I cannot leave Szechuan (Sichuan) and our never to be forgotten stay of two years at Tangchiato without recording a few stories which we love to tell, some of which we cherish. We walked frequently outside the installation and our favourite walk was down through the Brass Gong Gorge. Unlike most other gorges downstream, the sides of this gorge were not steep and it was possible to have a defined, though somewhat rough, tow-path. One day we encountered a 100 foot junk being dragged upstream by trackers. These fellows are amazingly agile, almost bent double with a tow-rope of woven bamboo strips over their shoulder and bellowing in unison. This operation requires unbelievable coordination with the junk-master on board steering and the laodah, head of all the crew, beating on a drum a rhythm of instruction to the trackers. We stepped aside on the tow-path to make way for the trackers. To our surprise and embarrassment, one of them looked up and smacked Rena on the bottom. It would not have been difficult to upset the entire tracking arrangement and imperil the junk battling its way against a pent-up force of the river, but we left it at that, only now placing the incident on record as our Brass Gong Gorge experience.

When it comes to looking at the skills of tradesmen it was not easy to draw comparisons, but among those we had at the installation one was truly outstanding. He was employed by Shell as a carpenter, mainly to handle packing-cases and minor woodwork jobs. One day, while in his tiny work-shop, I noticed an extraordinary number of tools for woodwork. Imagine my astonishment when on enquiry it was revealed that he was really a carver of idols for the temple. And we employed him, for the most part, to nail up and dismantle packing-cases. I have to admit, thereafter, that when he had spare time he did some wonderful work for the Blacks, using all his tools, figures from the Chinese Epic *The Three Kingdoms*, dragon carvings four feet high, from a solid block of wood, for standard lamps, and a set of tables very much prized in our home now. As a craftsman, using wood, his work was of the highest quality. Fourteen years later, on a visit to Chungking (Chongqing), I asked to see him, but he had passed on and his son told me he was incapable of following in his father's footsteps.

The kind of food that was normally part of our diet, particularly with two wee ones to cater for, was a problem. We ordered, in fairly large quantities, from Shanghai when our ships were running in the high-water season. The majority of our requirements were tinned and my next story is a canned one. As first, as we emptied one after the other, it seemed to us that the only place for the empties was the rubbish dump. Our cook, who saw what was going on, simply shook his head but finally made it clear to my wife he knew a better way of handling the empties. We had a back door leading to a path along which the country folk went to market. Our cook arranged all the empties outside of this door on the edge of the path. Lo and behold, to our surprise and delight, the country folk carrying their produce to the market stopped, picked up the empties and left in their place the loveliest vegetables, sometimes fruit. This was an area where the best of oranges were grown. They needed tins, big and small, and were prepared to pay in kind. No wonder in later years that my wife said it was while at Tangchiato we started to save money.

At the Shell Installation, Tangchiato, our life was dominated by the great river Yangtse. We lived about 300 feet above its low level, in a sheltered bay, where the river made a right-hand turn into the restriction of the Brass Gong Gorge. From an examination of the installation site after we arrived, it was obvious that in the planning stage provision was made for a rise in water of over 100 feet. The past revealed that such a volume of water came from the melting snows and monsoon rains at source. Although a rise in the river was expected, the phenomenal extent and its speed was almost frightening. One night in early July we were all bedded down comfortably when I was awakened by a sound like thunder, but it was sustained and steady. The river was on the rise and the rumble came from the restriction as it entered the gorge. For the next two days the rise averaged two feet an hour. The outlook over the countryside, now inundated with water, completely changed and wooden houses could be seen floating by. The suddenness of all these sounds, sights and happenings left an unforgettable impression on me.

This is a creepy-crawly story about a certain spider, the smallest creature imaginable. One night we had one in our bed and it laid a trail of poison on part of Rena's face and neck. To trace the spider with a view to elimination was difficult. It returned the second night, did more damage and was never seen again. The drag it left behind caused much pain and disfigurement. So beware of the Yangtse spider.

The near-by village of Tangchiato had two local industries which we visited on occasions because they were essential to village life and both products were exported world-wide. The Tung tse tree (*Aleuvites Cordata*) is at its best in Eastern Szechuan (Sichuan) and produces a nut, the oil from which when crushed is the base of the best varnishes and paints. The village industry obtains the oil by pressing the nuts already put together in the shape of a cylinder and assembled in a press. Thereafter the oil is squeezed out of the nuts by pressure using wedges, hand-driven by swinging a battering ram. Quite an exercise! The leaves of the mulberry tree are well-known as the food of the silk-worm which produces the famous Chinese silk. The highest quality was woven in our village.

We could not leave our Szechuan (Sichuan) home without a word on the garden. The climate was never really cold; frost and snow were unknown and rainfall was adequate, although strange, to say the sun was rarely seen between November and March. Be that as it may, we enjoyed a beautiful garden and although much of it was on a steep incline on both sides of the path and steps to the house, the soil was rich enough to promote the increase. Our gardener, Lao ho, needed prodding but he kept his domain tidy. We enjoyed his antics as the blossom was falling, when much sweeping-up made him pause from time to time, look up at the 'offending' tree and shake his fist. Attractive mulberry bushes, prunas and other blossom trees, as well as a great variety of flowers, made for delightful surroundings. Topping the flowers, our chrysanthemums would have made the Japanese envious. Vegetables were so plentiful from the hinterland that it was decided to grow only corn on the cob, principally for our chickens. This was dried on the tennis-court and stored for round-the-year use.

The Asiatic Petroleum Company's ship *Tien Kwang* was our transport over 2,500 miles of river to the coast. Journeying westward, against the flow of the Yangtse, we were over one month getting to Tangchiato. In the opposite direction, with current and mid-water-level conditions in our favour, we reached Shanghai in ten days. The gorge passage was fascinating but much quicker on the bosom of an eight knot current.

In early March 1937 we left Shanghai by P & O liner for London. The voyage was uneventful and we decided en route to leave the ship at Marseilles and travel overland. Our two daughters, aged three and four were looking forward to seeing grannies, grandpas, uncles and aunts etc., and we had a big welcome at Glasgow Central Station. I remember my Father, who was not to be long with us, welcoming 'June the Prune'. We stayed at Beechwood, the home of Rena's parents, from where Grandpa Yuille enjoyed taking the girls out for a walk except on occasions when Mother dressed them up in uniforms which he thought was somewhat on the sparse side.

During a visit to Uncle Willie, my brother, who was a schoolmaster at Drumore, Wigtonshire, we heard the sad news of my Father's passing on 1 May, so we had to return to Dumbarton. Father had a heart problem, in my view brought about through many hours of overtime working during The First World War. His job involved the fitting of turbine blades in casings and rotors, very often working in a bent-up position for hours. His working life with one company was just on fifty years and he stopped work on doctors' instructions. No attempt was made to recompense him for long service or to say *au revoir*. The company actually laid their hands on his tools which he had designed and made himself for the job. Father was sixty-five when he died and Mother, two years older, carried on to make the century. As a family, we stayed with Rena's parents and afterwards with my Mother at Hartfield, before leaving again for the Far East.

4

Price's Candle Factory and Tientsin

Time passed quickly. Shell's plan for me was right out of oil into their newly acquired Price's Candle Factory in Shanghai. Paraffin wax was a Shell product and practically all candles, even for church purposes, were of this raw material. Our leave was prolonged because of my stay getting to know something about the gentle art of candle making, not only for illumination but decoration as well. Nowhere was this art carried out in more detail than at Price's Factory in Battersea just south of the river in London. It was a complete change of atmosphere for me but interesting, especially as I was shortly to take over the factory in Shanghai where the production at peak was a million candles a day — but more of this later.

Rena and I managed to get a short holiday at Kandersteg in Switzerland where we made our acquaintance with skiing. We enjoyed relaxation in the Bernese Oberland where the weather turned out just right for fun in the snow and the hotel accommodation was excellent. Unfortunately, Shell advised me I would have to return to Shanghai alone. This was due to unsettled conditions there where the Japanese were bent on interference in Chinese affairs. Their immediate object was to take over the Chinese area which surrounded the international and French concessions in Shanghai. This was the beginning of a period of eight years of Japanese occupation of large areas of China, which in itself is very much part of my life for the same period.

A glance at the possible routes to the extreme Orient on a global presentation gives the impression that the journey via the Mediterranean and the Suez Canal is a long, long trail of winding whereas there is something attractive about the longer way there via Canada and the wide expanse of the Atlantic and Pacific Oceans. I chose the latter route for my return journey alone to Shanghai, all the way afloat or by rail, using the facilities of Canadian Pacific Railway Company, literally from door to door. I sailed from the Tail o' the Bank off Greenock, within sight of Dumbarton, by CPR ship for Montreal, then straight off the ship on board a train to see Canada for the first time, prairies and all. The attraction of the Rockies and that gem Lake Louise could not be resisted for a few days. Soon it was Vancouver, my favourite Canadian city, and once more afloat, on one of CPR's Empress ships, via Hawaii and across the date line to East Asia. I really did enjoy Honolulu, spending almost all day mastering the surf-board, tired at the end of it but with the knowledge that surfing takes a lot of beating once you know how. Long and beautiful days were spent crossing the Pacific by the great circle route, almost skirting the Aleutian Islands. Shanghai at last and down to earth at Price's Candle Factory, Robinson Road, in spring of 1938.

What a change in comparison with an oil installation! I was in at the deep end and was going to miss being with Rena and the wee ones very much. In the first place, the factory site was in a very down at heels part of Shanghai, on the banks of the Soochow Creek, the boundary between the international concession and the Chinese built-up area. The creek was an asset for transportation but as a boundary, with the existing political tension between Japan and China, it was dynamite. The accommodation at the factory was poor compared with Shell elsewhere.

The political situation was tense and remained so for nearly a year. The Japanese were engaged in gun-battles with the Chinese soldiers on the other side of the Soochow Creek. In effect this meant that refugees were leaving in their thousands to what was for them the safety of the international concession. Many were wounded and in such a poor state of health that they lay dying close to our boundary wall. Disease was rampant and we were afraid of cholera. Truly, Price's Candle Factory was in the front line. The so-called international settlement had a number of foreign troops at their disposal and the factory was so exposed to fighting that a platoon of US Marines was stationed on our site. What in all the earth they were expected to do I never did find out. Just their token presence, I suppose, was enough to keep intruders out. I remember one day we all came under some heavy fire and I have never seen anyone take cover so quickly.

How did we justify our existence in turning paraffin wax, with a column of wick, into candles? In the first place, as the political situation developed, the demand hit an all-time high in a very wide area of the hinterland. This was simply due to the fact that it was impossible to transport kerosene in cans. The need for illumination in a wide area was left to the humble candle. Clearly this called for diesel electric plants but so little had been done in this connection that the candle was the only means the country folk had to illuminate after dark. Our candle factory worked for long spells, day and night, seven days a week. We produced 1,000,000, twelve ounce packets of six candles every day, labelled and boxed, in terms of weight, over fifty tons of raw material every day. The heavy work was a man's job but women did all the packing. In total we employed about 300 people. In supervision, I had to call on the main Shell office for assistance and my wife made many additional meals daily, starting from breakfast.

Rena and the family joined me after six months and we took up residence in the manager's house. Our social life was curtailed a little due to pressure of work. Fiona picked up a spotted complaint and nearly had to leave the ship in Hong Kong. Shortly after arrival, school was very much part of the family for the first time. I well recall an incident worth repeating: Fiona, a year the senior, was prepared for her first day at the Cathedral School. June would be next year. But it did not work that way as June said, 'I am going too.' And she did! Although somewhat removed from the centre of things, transportation was not a problem. We acquired a Skoda car and were mobile thereafter. In the early summer of 1938 we were able to go for two weeks to Tsingtao (Qingdao), a delightful holiday resort with splendid beaches on the Yellow Sea and partially sheltered by the Korean peninsula. We enjoyed this vacation in the company of friends. Many White Russians had settled there so there was a variety of food, with some dishes outstanding.

Cheek by jowl with our candlemaking, we had two never to be forgotten Soochow Creek neighbours who should be mentioned before we take off for pastures new. The gentle art of making silk and the Chinese part in it originally needed no emphasis. Suffice it to say that mulberry leaves and the silkworm are synonymous. The process next door to us was one of making sure that the worms had all the mulberry they could comfortably digest. It was, however, the next stage that upset us. As the worms shed their cocoons, the silky protective envelope which they secreted resulted in giving off such an obnoxious smell that even the Chinese candle-moulders could not take it and a fall in output was the result.

Next but one to us was quite a different operation. A brand new incinerating plant from Germany had just been installed to take care of Shanghai's tremendous volume of rubbish. The machinery was the very last word in not only incinerating but in separating secondary products. The German engineer who supervised the working of this juggernaut came to see me one day with tears running down his cheeks. 'How is it possible,' he said, 'that there is absolutely nothing combustible in their rubbish? It will not ignite — what am I to do?' He found out the hard way that the Chinese never dump any thing that is combustible. It is part of their economy that anything can be used in the kitchen to fire their cooking stoves.

Before leaving for my next appointment we had an opportunity of a short holiday to the Liao-tung (Liaodong) peninsula protruding into the Yellow Sea from Manchuria, just north of Shantung (Shandong) Province. Port Arthur was the name given to the town by the Russians when they annexed part of the peninsula from China late last century. The name of the principal town then changed to Dairen. Back in Chinese territory after the Second World War, Dairen became Luda and what was formerly Port Arthur is now Lushui. After all this geographical explanation, what happened to the Black family? It was the month of September 1939. Should we go to Dairen or not? We did go and had anything but a pleasant holiday. The whole area, the Peninsula of Liao-tung, was under Japanese control, as were all the provinces of Manchuria. Then a real comic situation existed in the latter; no one but the Japanese could have conjured up such a ridiculous situation. They took the last Ch'ing Emperor, Pu Yi, who was living in obscurity in Peking (Beijing), and made him Emperor of Manchukuo. Enough comedy on the puppet, Pu Yi. The Japanese were two years away from declaring war, but they were moving rapidly towards it. Our holiday was with all this around us: curfew at night and illumination black-outs. One night we had a rat in the bedroom and I switched on the light to get it. All hell was let loose: Japanese gendarmes were shouting and doors were being thumped, so all in all we were glad to cut our holiday short and return to Shanghai, where we arrived in time to hear Neville Chamberlain's declaration of war on Germany.

I was off a few days later to Tientsin (Tianjin) in North China by coastal steamer. Rena and the girls were left behind in Shanghai. This time floods prevented them joining me for a few weeks. The oil installation there was extensive and on the banks of the Pei Ho close to town, opposite but outside of the British concession. The concession here was long established with lay-out and buildings well planned in true Anglo-Saxon style. A very awkward situation existed for anyone living outside and this applied to the installation staff who were housed on site. The Japanese had ringed the British concession with their troops. In effect it was a blockade, which was very provoking, as was intended. The Shell office was inside, as was school, church and social life. After all, we were not at war but their antics were most irritating.

The Shell plant, with can-making machinery and a bulk storage capacity of considerable volume and variation of product, which was topped up by lighter from Tangku (Tonggu), at the mouth of the Pei Ho, was my responsibility. All outgoing products were despatched by road, rail and river, some in bulk but for the most part-packed. Problems involved arising out of the Japanese blockade were always in mind, as very much so was the war in Europe and its possible extension.

Life was not a bed of roses at that time in Tientsin (Tianjin). There was, however, something to be said for living outside the blockaded British concession. Marketing and freedom of movement were on our side. As school was inside, this meant entry and exit the same day. The girls were then five and six years old and Mother had to accompany them. The Japanese were always polite, especially to the wee ones. Their English language was almost non-existent but they had become accustomed to saying to the little girls, 'I lova you' and Fiona said, when at home, 'What should we say in reply?' I told them reply without hesitation, 'My daddy lova you.' The Japanese checked passports and would not allow anything to be taken in to the concession. Trucks could be held up for days and in consequence food stuffs were ruined. The Blacks had to come through the check point for church on Sundays. This was a new one and called for the officer in charge who said brusquely, 'No school today. Where are you going?' When he heard, 'To church!' all obstacles were put aside and we were given the Japanese equivalent of a big hand. Treatment at the check through stage could be difficult, indeed from time to time men were ordered to take off their shoes and socks and walk over lighted cigarettes. We employed a Japanese to assist with translation problems. Poor chap, he was regarded as a traitor by his fellow countrymen.

Before leaving this period of Japanese-occupied-China in the late 1930s, I would like to go back to a worthwhile story about Miao, my old Chinese teacher when I lived along the Yangtse at Chinkiang (Zhenjiang). From time to time I talked to him about getting married but never seemed to get very far. It was nine years later, just before I left Shanghai for Tientsin (Tianjin), that our paths crossed and I raised the point of marriage. To my astonishment he said, 'I am married and this is how it came about. A friend of mine, who knew I was looking for a wife, informed me one day that he was in touch with a rich fellow countryman who had a concubine and was forced, through difficulties of Japanese occupation, to bring his entire household to live in a hotel in Shanghai. In somewhat cramped conditions, his entire concubinage was more than he could bear. For this reason he made it known to his friends that he wanted to get rid of several courtesans. I decided to look them over and made a choice of one, on the understanding that we would live together for a year. If at the end of this time we were both happy with our way of living then I would legalize the arrangement. In conversation after six months of our trial-get-together I made it clear I was unhappy with her ability as a cook and marketing expenses kept on rising. Consequently if, by the end of the year, there was no improvement we would part company. After this straight from the shoulder talk her whole approach to housekeeping changed, so much so that not only did I save money on marketing but the food, as we Chinese say, was *hen hao chih*, very good to eat.' My reaction was immediate. 'I want to meet this girl!' Her cooking was superb. The only thing she could not understand about me was that I liked soup served first and she insisted that it should be served, in accordance with Chinese custom, as the last dish of all, a mere bagatelle after such a wonderful meal.

Years later, in 1949 to be precise, I met Miao in Shanghai. The communists were advancing on the city and I asked him what his plans were. He replied, 'I must leave Shanghai and go south.' I replied, 'Don't leave it too late, you might not be able to take your wife.' His answer was immediate: 'If necessary I will carry her.' Some time later we met them both in Hong Kong. Miao stayed with us for two weeks. He was a typical North China type. Hong Kong for him was a foreign country. He could not speak Cantonese, but he was soon in the business of making rattan furniture and exporting it to America.

While in Tientsin (Tianjin) we had the opportunity of going to Peking (Beijing), about 100 miles west. Such a break to see the Imperial Grandeur of this unique city could not be missed. We were warned of poor road conditions. This turned out to be only too true, giving the Skoda and its occupants a pounding; so much so it was decided there and then for the family to return by train. But it is to fascinating Peking (Beijing) we now turn our attention. As a capital city it is not very old, sometimes one hears it talked about as the ancient capital. In point of fact, as a capital city, it only came into prominence when the Mongols conquered China and Kublai Khan was the first Emperor of the Yuan Dynasty in 1280. But it is to the Ming Dynasty, 1368 to 1644, particularly to the Emperor Yung Lo for devising such a lay-out and bringing his architectural plans to fruition, that makes Peking (Beijing) a capital city all on its own. Much has been written, with illustrations, about Peking (Beijing) and this brief account does not aim to take the place of a guide-book. Suffice it to emphasize a few places of outstanding attraction. The centre is undoubtedly the Forbidden City, with its mass of glazed shining roof of yellow tiles surrounded by yet another square of pink crenellated walls which mark the Imperial City. Peking is a city of temples, the most impressive of which is the Temple of Heaven with its gilt ball at the top and below it a triple-roofed azure-tiled gold-capped shrine. I recall that as we entered the white circles of marble balustrades which support and elevate the whole structure, we all gazed in awe at its beauty and it fell to Father simply to say, 'This is the Temple of Heaven' which was followed by an immediate response from June, then six years old, 'I thought Heaven was only in Scotland!'

A visit to Peking (Beijing) would not be complete without making one's way in a leisurely fashion through the markets and shops where durable art forms are on sale. We only made two purchases, both still in our possession. Cloisonné ware was an attraction. This art of forming cloisons to hold colourful enamel was a decorative art developed in China 'away back'. It was, however, believe it or not, an impetus from Byzantium over 1,000 years ago which enhanced long-developed Chinese skills in this art. Cloisonné held our interest and we purchased a set with the colour green predominating in the design in all four pieces. Alas! The loveliest piece of all, a beautiful plate, makes the set incomplete. Finally, in a bookshop I obtained two volumes, in English, which have since been invaluable to me, *The Ruins of Desert Cathay* by Aurel Stein, published in 1912.

A glance at the map of China will show that the Desert of Gobi is not far from the area where we lived in the province of Hopei (Hebet). The word 'Gobi' is of Mongolian origin and describes the nature or type of desert and has no connection with its geographical spread. This great desert played not a little part in our lives in Tientsin (Tianjin). For a period in the autumn of each year, though its intensity covered only a few days at a time, the wind from the Gobi picked up the fine sand and carried it for 100 miles or more. It was air-borne for just long enough to fall on the Tientsin (Tianjin)/Peking (Beijing) area. To combat this plague of sand, all windows were double-glazed long before we in the west found application for it. But so fine was the sand it was virtually impossible to keep out of the houses; even food in the 'fridge' was not immune.

Under the circumstances prevailing in Tientsin (Tianjin), it was a relief to get on a train for the four hour's journey to Pei tai ho (Beidai he) on the Gulf of Chihli, now named Bo Hai. The last word means 'sea'. In 1940 the family were able to stay there for much of the summer, with visits from me at the weekends. Pei tai ho (Beidai he) was a gem of a place after the rough and tumble of Tientsin (Tianjin). Our rented bungalow was quiet and close by the beach where the girls made their acquaintance with swimming. Our mongrel dog Abdull proved to be a first-class guard dog. Even at the weekends he was reluctant to let me in. Unfortunately, he developed an anti-stranger tendency and we had to have him put down.

Shanhaikuan (Shanhaiguon), within a stone's throw from Pei tai ho (Beidai he), is at the eastern end of the Great Wall, whose length of over 2,000 miles terminates in the foot hills of the Tibetan Alps. We visited this historic site and saw something of the zigzagging wall, about which I will write more later.

All our oil-in-bulk supplies came by lighter from the port of Tangku (Tanggu). The tanker terminal was at the mouth of the Pei Ho. The last word means 'river'. In an examination of a lighter that had been discharged, I failed to get a foothold on the ladder below the hatch-cover and fell some seven feet to the tank bottom, striking my head on the way down. At the time there was little or no after-effect, but later and for years after I was subject to dizziness and fainting in crowds, even in small numbers. People were anathema to me. Be it in church, cinema or even a bus I was overcome by the same feeling. A number of medicos in their examination failed to get at the trouble. I would say this state of affairs lasted for more than ten years. Happily it disappeared, but while it lasted it was very disquieting, not only for me but Rena too.

5

Australia and Return to East Asia

And now for a big change in environment. It was the Spring of 1941 and we were due long leave. Europe, at war, was out of the question, so we set our sights on Australia. I was not sorry to leave Tientsin (Tianjin), where signs of things to come were close at hand. We sailed by coastal steamer for Shanghai from where, after a few days, another coaster took us to Hong Kong. Joining a much bigger vessel there, whose chief officer was a Dumbarton man, we set sail for Sydney or Melbourne via the Phillipines. In Manila an old Shell colleague came on board and we lunched at the Army and Navy Club. From Manila, we skirted the many islands of the Phillipines to Zamboanga at the southern tip of the Island of Mindanao, where, it is said the monkeys have no tails. By the way, we did not see any. By now we are well south and approaching the doldrums and then that mysterious line, the Equator, where Father Neptune takes to the swimming pool with the passengers, who are faced with walking a greasy pole. Every day it was getting warmer as the ship threaded its way through the islands of Indonesia, with New Guinea not far off on the port side. Our passage was inside the 1,250 miles Great Barrier Reef. By this time we had decided to disembark at Sydney. Apart from an occasional blackout on board ship, as an exercise, it was a calm and enjoyable voyage, but there was in the atmosphere a threat of what lay ahead.

We arrived in Sydney safely and the girls were astonished to see so-called 'white men' doing the work of longshoremen or dockers. To them this was a job for the locals and, to date, their locals were all from East Asia. Our ship berthed in the shadow of Sydney Bridge right in the heart of this wonderful expanse of harbour. The hotel was in a perfect setting at the entrance to the harbour known as The Heads. At the South Head, just a step further than Vaucluse Bay and inside the Heads, was a lovely situation to observe shipping. A sight we will always remember was the giant Cunarders, *Queen Mary* and *Queen Elizabeth* passing at The Heads, one full of troops en route to Europe and the other coming in empty to load more boys for war. It was not to be long before the passage of Australian troops on the *Queens* would be in the reverse, to defend the homeland. The food in our hotel was excellent. We especially enjoyed afternoon tea with scones, cream and jam. The girls brought us to heel very sharply. We tried in vain to get them to go down to breakfast before us. After the first morning we failed and were faced with the excuse, 'We don't like the waiters, they are too cheeky.' Being used to oriental service, which is good and very polite, almost too prim and proper, they could not understand what seemed to them the bold approach of the Australian waiters.

We were now mid-way through 1941. We moved from our very pleasant hotel at The Heads to a rented flat in North Bondi overlooking the much talked of Bondi Beach and near-by golf course. Both of these facilities were used and enjoyed. Bondi was famous for its surfing, and as much of the golf course fairways were along miles of cliff, with a sharp drop to the sea, many balls were sliced into the ocean. It was a restful time for us, with many visits to other parts of Sydney. I had an appointment with a doctor in Macquarrie Street to check up on the happening resulting from the bulk-lighter fall in Tientsin (Tianjian). All he could say was prompted by the sight of my tartan braces, which elicited the remark that I carried my natural tendency or maybe my identity to quite a length. However, he did assure Rena I would be around for quite a while yet.

With travel documents to Melbourne still in our possession we decided to make the trip by sea and stay there briefly. In order to be sure of occupancy in our Bondi flat on return, we paid rent in absence. Little did we know our landlady had found a convenient tenant in our absence but forgot to straighten out the bedrooms for our return. Back in Sydney, our next visit out of town was to the Blue Mountains, quite a change and to our liking. The hotel, in quiet wooded country at Leura, was lovely. Breakfasts will always be a milestone here. Hearty thick lamb chops were the mainstay on the menu. Regrettably, this was beyond our appetite so early in the day. In the valley not far from the hotel we enjoyed horse-riding. For me in the years ahead it was pleasant to look back to these days in the Blue Mountains of New South Wales.

The experience of being in Australia, even for the short time of my stay, brought me face to face with the Aussie point of view. This is what I mean! While in Sydney there was much ado in the press about the surplus stock of wine in the country. With all this talk in mind and thinking how fitting it would be for me to help reduce the surplus, I called on Penfolds — the name is synonymous with wines and spirits in Australia — to enquire about wines. Imagine my surprise when the man behind the counter, looking at me with some astonishment, said, 'Nobody drinks wines here. This is a beer-drinking country.' Just recently we had a visit from an Australian friend on holiday here and she made it quite clear that in present-day Australia those who can prove they have a relationship with Botany Bay are very *persona grata*. Another Aussie story worth telling comes from a fellow passenger in a bus. He could see I was a stranger in these parts and felt the occasion was ripe to put me on the right lines. This he did by making it clear that I would get on fine 'down-under' as long as I did not think all Australians had a Botany Bay connection.

My wife and two young daughters, who had four years in Australia on their own, were grateful for kind offers of assistance and the many friends they cultivated.

And now, for me at any rate, came the last lap in Australia. Late in August, 1942, on a visit to Shell's office in Sydney, they informed me there was a message from Shanghai about my return to China. In effect, though, the underlying meaning of that message was quite clear. They had no objection to all of us coming back together but it was advisable in the meantime for the family to stay in Australia. I booked a passage to Hong Kong and on to Shanghai on my own. At the parting it was a sad farewell and, although we did not know it then, destined to be a long one. It was, however, wise for Rena and the wee ones to stay in Australia. The war in Europe was building-up and clouds were definitely appearing in our working part of the world, North China. The Japanese were moving further inland in China and taking control of the coastal strip south of Shanghai. The return passage to Shanghai was without incident, though the wife of the chief accountant and part of her family came on board at Brisbane. I am sure she regretted this later, but that is another story.

In late September 1941 I had returned to Shanghai and was appointed to a new job in head office. The senior marine staff had been called to London and I was appointed to take charge of the remaining coastal ships and the movement of small craft on the rivers. It was new and interesting work and brought back early days on the China Coast. Tankers of 10,000 d.w.t. were not coming north of Hong Kong and we only had one ship of our own bringing gasoline from the Dutch East Indies to Shanghai. Stocks were low but always sufficient to keep road transport on the move in Shanghai and surrounding areas. My living accommodation was a flat above the office, which reduced transport problems. As it transpired later, being in the centre of things had both advantages and the reverse. My next-door neighbour and fellow engineer had a Chinese girl friend. They were both keen radio enthusiasts and had equipment in their flat capable of transmitting and receiving messages internationally.

The policy in Japan towards China then was one of territorial encroachment. the coast was of prime importance to them. After occupation of ports, they pushed inland. This was especially true in North China and the Yangtse delta area. Already well-established in Manchuria, Inner Mongolia was ripe for picking, as were the coastal provinces of Hopei (Hebei) and Shantung -d3-Shandong). No outside power was prepared to dispute Japanese movements in China, understandably so, as the war in Europe, by late 1941, had cast its shadow world-wide. Also the Chinese government then were incapable of resisting. By this time, too, Japanese propaganda continually referred to their affinity with what they were pleased to call their 'Greater East Asia Coprosperity Sphere'. In early December 1941 it was evident to anyone living in Shanghai, surrounded by the fire power of Japan, ashore and afloat, that the future was bleak. The foreign staff of Shell in the North China area was approximately fifty, for the most part with families. Approaching this late hour, some thirty, a few with families, remained. A notable exception was the General Manager who made a getaway just in time. The preceding brief account sums up the position in China, as seen from Shanghai, early in December 1941. On the 7 December 1941, the Japanese attacked Pearl Harbour, Hawaii. This for us in occupied China was fat in the fire. In faraway Sydney, Australia, North Bondi to be precise, the Black family were wondering what was happening to Father. In the wider sphere 'down under', the Aussies were beginning to realize, for the first time, that they were not too far away from the land of the rising sun.

As expected, from this time on, the pattern of our lives in Shell changed drastically. In point of fact, it was a year before we were actually confined to quarters within a barbed-wire fence and sentries on the perimeter. Meantime, soldiers were on guard duty outside head office and, inside, officers in charge of a special military detachment were checking personnel and documents. There was no need for them to spend much time on marine movements. My work was more of a naval matter and, anyway, our tankers were long since out of their immediate orbit. At this stage, and in head office building where I lived, I was more concerned about my colleague next door. Attention has already been drawn to his radio enthusiasm and I was concerned about what the Japanese would find when they began searching his flat. I was ill at ease, not so much with my associate's radio hobby but with his up-and-coming Chinese girl-friend who appeared to me to have more than a passing interest in wireless. She was a smart young woman and in the next few hours I was to find out the way she channelled her intelligence.

6

Incarceration During War in the Pacific

On 12 December 1941 the Japanese incarcerators had turned their attention to the flats where some of the staff lived over the office. I was well aware of the fact that when they entered next door their eyes would boggle when they saw the extent of the radio equipment. What I did not know was how much the Chinese girl had been involved in the use of this equipment to send messages to her government in Nanking (Nanjing) and elsewhere about Japanese activities locally. In this connection, call signs and a record of the information transmitted had not been destroyed. After several hours of questioning, the Japanese interrogators locked up all entry to the flat, so they thought, saying they would return shortly. A communicating door between our two rooms was not locked and my colleague, in a panic, put me in the picture. In effect he said that, if they found the information still in his apartment, sent to stations in China through his equipment, he shuddered to think what their reaction would be. The information must be destroyed without delay, but how? The immediate reaction was to get it into the sanitary system, the loo. On second thoughts this idea was discarded because of our suspicion that the Japanese would be alert to such a possibility covering the entire building. Finally I agreed to stuff my person with the evidence and take it out of the building. As yet we were allowed to come and go through the main entrance but who could say when a search would be made by the soldiers on duty there? I decided to take a chance. It was winter and heavy clothing was in order. This helped to conceal the evidence. Fortunately, I made it, passed the armed guard, then spent hours walking through the streets of the French concession tearing up the proof and depositing it in rubbish drums streets apart.

The radio connection with my colleague just described resulted in a bond between us which lasted for ten years through thick and thin. For reasons never quite clear to us, the Japanese had been able to persuade Shell's deputy general manager to do a survey of the international concession, Shanghai, with a view to locating stocks of fuel oil. Presumably a certain pressure was applied and, being at war, he was in no position to do any other. This is how we became involved in tracking down stocks of fuel, above and below ground, in the labyrinth city of Shanghai that were of value to the enemy. Afterwards, having been ousted from our apartment above the office, we stayed at the home of the general manager and fitted in well with the scheme of things prior to being gathered into the Japanese net and locked up.

As a family we had a number of Chinese friends in Shanghai, in particular my old language teachers Miao and the Zings. The latter was a marriage of east and west which, despite the great cultural difference between the parties, was a success. He was in the textile business and had an early training in Blackburn where he met his future wife, 'a lassie from Lancashire' who, through the church in Shanghai, became very friendly with my wife. I had the opportunity to see the Zings on several occasions during 1941 and left in their safe keeping some of our treasures. Before segregation from the Chinese and so-called neutrals of Shanghai, we moved house to Rue Ratard, Shell's residential estate. This was due to seizure of the general manager's house on his departure behind closed doors.

In the penultimate paragraph mention was made of the bond between us arising out of the radio connection hence my use of the personal pronouns 'we' and 'our'. I used this link, which was applicable during the lengthy period of Japanese incarceration when we had the opportunity to plan certain further radio connections.

A guest at our final halt in Rue Ratard was the manager of Kelly and Walsh, booksellers and publishers in the Far East. He was an excitable fellow, as this story reveals. One day he returned from his beautiful book shop in Nanking Road to let us know the end had come for him. The Japanese were going through his stock and were about to lay their hands on a super

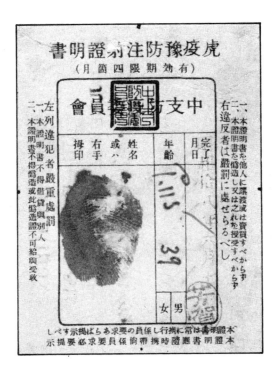

The author's thumb mark taken by Japanese for identification

publication of Madame Butterfly in story and music. It was well-known to us that in Dai Nippon they regarded this as a slur on their culture. He appealed to us for help — what should he do? My reaction was, in effect, that if you can get the book here before the Japs find it I will take it and give you in exchange a cloisonné plate in green glaze. My family have never forgiven me for this swap. They were not Gilbert and Sullivan fans and really loved this beautiful piece of cloisonné enamel ware.

A British Residents Association was formed in the early days after 11 December 1941 and the Japanese issued their instructions to the British, through this body. We were nearing the end of 1942 and they were at last organized for segregation, date, time and assembly point together with what was allowed to us as baggage. At this point in time my day to day notes take over.

From then on, for what seemed ages to me, I was known to the Japanese as P115, aged thirty-nine — my identity slip of paper is attached hereto. With the passing of time the back of this slip was used to establish the fact that injections against cholera, smallpox, typhoid and para, A & B were given to P115.

Prologue

Four hundred of us were notified on 24 January that we had to prepare in a week's time to depart the Shanghai life for Pootung 'more far'. Instructions were issued regarding baggage which had to be carried. Fortunately, an iron bedstead was allowed and I decided to take a camp-bed and bedding as well as the old four-poster. Speculation was rife about the place of our detention in Pootung, but no official news was ever given as to the whereabouts before our departure from Shanghai. After a week of hectic buying by way of wearing apparel, provisions and gadgets, much of which proved useful later. At the time of these whirlwind shopping expeditions we were impressed by the Chinese shop assistants who were conscious of what lay ahead. They never failed in their courtesy and service to the 'foreign devil'. Everything from a boiler-suit to a screwdriver or a tin of sausages was in demand. It was a heterogeneous collection of goods and clothes enough to stagger the average individual in ordinary times. In addition to all the hectic buying, arrangements were made with neutral friends to keep valuables and to leave money with them for the rainy day when a parcel would be manna from heaven. The BRA functioned at high pressure, looking after all details big and small for the so-called comfort of the first batch of internees. They did a good job and it is to be hoped though it hardly seems likely that a skeleton staff will be able to operate at Church House throughout. The British community has now come to the conclusion that practically all will be under lock and key in the next few weeks. Most of us were on 'top of the world', although there were doubts, among those with children, about how they would fare under cramped quarters and what would undoubtedly be trying conditions.

January 31st 1943

On the day of assembly it was laid down only such baggage that could be carried by the internee himself, would be allowed. I was overloaded but took a chance that later led to a few anxious moments. Those officiating acted tough and in some cases refused to allow outside help with baggage. The BRA had engaged transport for packages. This assistance was vetoed and the luggage forcibly removed from hired vehicles. The assembly was timed for ten am but it was one o'clock before my section, twenty of us, moved off. Practically all of us were weighed down with impedimenta and we literally staggered to the Bund from Church House where we said goodbye to friends. It was noticed that the general arrangements were in the hands of our residents' association. A Japanese presence seemed necessary only at the final roll-call. A crowd assembled near the Custom's Jetty to see the first British and Americans off to internment. Demonstrations were forbidden and the tender was soon off upstream across the River Whangpu.

At last we knew something of our future home. Embarking almost opposite the French Bund we had a trek of a quarter of a mile to a large godown or warehouse of the British American Tobacco Company. With more time at this stage of the journey, baggage handling was not a problem, as there was no objection to frequent rests. The godown accommodation arranged for 400 of us was ill-prepared, in every respect. The place was filthy, sanitary arrangements were incomplete and inadequately planned.

We were escorted to our new place of abode by sentries of the naval landing party who provided the guard for the outer limits of the camp.

A Japanese civilian was camp commandant while his aides were consular police. The latter conducted all roll-calls and policed the inside of the camp. An oath drawn up in three parts was handed to us, for signature, on arrival. In effect we were under oath not to attempt to escape. And in a short speech by the Commandant we were forcibly reminded that any attempt to make a getaway would endanger our lives as the sentries had instructions to shoot on sight. Having shortly before this referred to his charge as a Civil Assembly Centre, these final words sound very hypocritical in comparison. The inauguration ceremony completed, we left the dining hall to look for a place to put our bed. Fortunately, I was able to get a place at the window in the middle floor of this tobacco warehouse. The building was 'T' shaped and faced west, overlooking the tall buildings of the Bund on the Shanghai waterfront. The Custom's Clock would be our time-piece in the months, possibly years, to come.

The workmen had just left the place allotted to us. The floor was in a dreadful state. Every time anything was moved a cloud of dust enveloped the package. We set to with a will and put beds in place. My allocation of space was sixty square feet for all purposes, later reduced to fifty on arrival of the next batch of the Emperor's 'guests'. The first meal was at four in the afternoon. Having eaten nothing since breakfast time, we were glad of it. The meal was a plain stew, an indication of what might be expected from now on. The area of floor space occupied by our section had a capacity for eighty men. Six small stoves, suitably situated, provided the essential heat, otherwise we would surely have frozen. It was following much speculation and some planning for the immediate future that we completed the first day's internment before getting into bed for a well-earned rest.

February 1st 1943

The camp routine was not strictly observed on the first day. We were all afoot early and after a breakfast of bread and tea pulled out overalls and set to with a will on the work of cleaning up. The floor had to be damped down and swept before mopping with water. Thereafter, the only way to tackle the job was to get down on hands and knees with a scrubber, soap and many cloths. Copious quantities of clean water were required. Those on the heavy work of swabbing kept at it while others supplied clean water. It was a back-breaking job and hard on the hands, using icy cold water. However, with the grain of wood on the floor 'coming up' on drying we felt amply rewarded.

February 2nd 1943

The daily routine as laid down by the 'sponsors' was as follows:

Lights: 7.00 am   Reveille: 7.30 am
Roll call: 8.00 am   Breakfast: 8.30am

Fatigue duties in the forenoon:

Mid-day meal: 12.30 pm   Supper: 6.00 pm
Roll call: 8.30 pm   Lights out: 10.00 pm

Jam was introduced at breakfast — one tin among five — otherwise the first meal of the day seems to have settled down to bread and tea. In the morning tea was sweetened but at other meals no sugar was evident from the taste and as for milk, it was conspicuous by its absence at all times. Herring was introduced at midday meal today, two for each man with a little cabbage and the inevitable rice. At last some order is coming out of the initial chaos. An endeavour is being made to find a place for everything and having found it to keep it so. Shelves and hooks are appearing as if from nowhere but thanks to shopping before we were swept into the net.

February 3rd 1943

An egg appeared today for breakfast. Looks like being the first and last. There are still many things to be done about tidiness. It is surprising how everyone is rallying to the cause of neatness, though some have little or no idea of what it means to be in this class. The sanitary arrangements are poor. Washhand basins are too small and WCs antiquated, while the overall water pressure is so low as to render sanitary and washing-up arrangements inadequate. So far there is no arrangement for bathing, though showers are in the offing. Today we received the regulations governing the Pootung Civil Assembly Centre. These comprised four articles: article one having eighteen clauses. In effect these regulations put us 'on the spot'. The food allowance having been laid down by the government, no alteration can be made nor will any complaint be allowed. It was with some amusement we read that 'The Civil Assembly Centre must be loved and cherished by all who live in it'. If you could see this place with barbed-wire, sentries and restricted space for exercise, the hypocrisy of this clause would be apparent. Sectional responsibility is such that if anyone makes a get-away, all in the section will be punished. All chores around the camp, with the exception of cooking, will be carried out by internees.

February 4th 1943

Assisted in scrubbing out dispensary. This room was made available for the doctor. It was makeshift and entirely without equipment. The camp organization is beginning to take shape. Police and fire squads were chosen, so many from each section. The former wore special armbands and were exempt from other duties. In our very mixed camp a police section is most essential. Other necessary committees were formed to run the camp at this time. 1,500 men in captivity, left to their own devices, within barbed wire could get up to all sorts of mischief.

February 5th 1943

The daily ration of food does not alter and is likely to remain as follows until the end of the show: breakfast: bread and tea with a little sugar. Mid-day meal: fish, or every other day a little stew with some vegetables, rice and tea with no sugar. Supper: same as mid-day. Often the meat or fish is absent, but when supplied, the weight in any one day, per person, never exceeds four ounces, including bone. Our daily meals are considerably supplemented by drawing from the stores which we brought into camp. When this supply is exhausted the ration provided by the Japanese will be very much short of requirements. A move has been initiated to deprive us of the money we brought in, despite the fact that prior to internment the Japanese stated that cash would not be confiscated. This will put 'paid' to any canteen suggestion.

February 6th 1943

The personal appearance of campers had now definitely taken shape. The sheep have been separated from the goats insofar as those who have decided to grow beards now look very hoary round the chin compared with their clean shaven comrades. It can be said that twenty-five per cent of the camp have decided to keep warm with a heavy growth.

February 7th 1943

Church Service got under way with a good attendance. Lack of hymn books and musical accompaniment was evident but those present expressed appreciation of the kind of service and sermon. Otherwise Sunday passed much the same as other days.

February 8th 1943

I do not subscribe to the general superstition which labels Monday as 'Black'. For me each day of the week is what you like to make it. However, on this particular Monday circumstances were beyond my control. The food for the midday meal did not arrive until two pm before a very hungry camp was summoned to eat. The straw that broke the camel's back was cuttlefish or 'squid'. Just to put an edge on our appetite we were informed that enough 'squid' had arrived for three days. Cuttlefish to most of us is to be found on beaches; there it is gathered by those who keep canaries as a bill conditioner. For the majority of us it was so tough as to be uneatable and being as it is a species of octopus it was repugnant.

February 10th 1943

There are certain jobs in the camp which keep one out of fatigues connected with garbage and disposal of sanitary effluent. The permanent jobs were police, kitchen and dining area. The mechanical side of the sanitary system was also in this category. The hot water problem, though not acute, was difficult. With hot water such a scarce commodity, a good and voluminous vacuum container is an asset. My flask capacity being considerable, I was able to loan to anyone in short supply. The kitchen was the only source and the hours restricted. So far the difficulty of obtaining hot water for washing and laundry still remains a 'can of worms' but it is on the way to being solved.

February 11th 1943

Our wedding anniversary has been celebrated by an official announcement from the 'powers' that they will in due course improve conditions in the camp. One may guess from such a statement that conditions were in need of improvement and go further in thinking our captors had no such intention. Indeed, their favourite expression at the moment was 'endure while we concentrate on fundamentals'. Evidence supports the belief that within the next few days many hundreds more will join us.

February 12th 1943

Cracked wheat porridge, so much enjoyed at breakfast yesterday, did not appear today. We wondered why! There must be a large proportion of campers of Scots' background judging by the accents of many around me.

February 13th 1943

The Chinese cooks, it would seem, have been used by some of us to contact outsiders. The Japanese are furious and have ruled that cooking will be carried out by the internees themselves. My first effort at physical exercise was strenuous. The class promises to be well worthwhile and I propose to keep it up. Today was 'shower' day. It is lovely, at long last, to get a fine hot 'spray'. On the scale, stripped, I am 145 lbs. The event of the day was the arrival of the baggage for a further contingent of internees. All of it had to be handled by those who knew their way around. This was a heavy job, as many stairs were involved in the handling of baggage for over 700 people. My first effort at physical exercises was tough going. The class started some days ago and the instructor has been stepping up the pace a bit, with the result I felt the strain not a little. However, the class promises to be very good for keeping fit, but much depends on the food ration. This evening's meal was the first cooked by the lads in the camp. If their preparation of rice and fish is indicative of their ability in the kitchen, it augurs well for the days ahead.

February 14th 1943

Heavy rain makes the arrival of bedding for the newcomers a sorry sight. Many of the mattresses will be unfit for use until they are well dried, which will take days. There would appear to be many more packages and bigger ones per head than when we came in. Obviously they are taking a chance, having heard of our short rations. This was a back breaking day, helping to move their heavy luggage almost solid with canned goods.

February 15th 1943

An important day in the life of the camp. Due to the heavy work anticipated with the new arrivals, breakfast was cut out. The first meal of the day was announced at eleven am. It was stew and tea. The beef was as hard as nails but the rice and sweet potato well edible. The first of the new arrivals 'rolled in' about noon. As with us, they were staggering under the weight of last minute packages. There are many elderly men in the contingent, a good number over sixty years. They will find it hard to 'stand up' to this tough life. In all 650 are expected, making a total of 1,000 in the camp. From the newcomers I was fortunate to receive some additional money, sugar and sweets from a Chinese friend and an apple tart from the Grants. From a casual glance at the new arrivals they are in the majority a well-seasoned crowd. The foreign population of Shanghai was nothing if not diverse, from the waterfront to the peak of managing multi-national companies, but here, in this so-called Civil Assembly Centre, 'we are a' Jock Thamson's bairns'.

I have just quoted from my diary the happenings in the first two weeks of internment to give an impression of the radical change involved in my life style. What follows is also recorded, but with emphasis on the highlights in the next few years when I was 'a guest of the Emperor of Japan'.

The crew of the American liner *President Harrison*, captured by the Japanese at sea after declaration of war, were among the new arrivals. Many were Negroes, including Bill Hegeman, the leader of the jazz band on board ship. They were a tough outfit but a great need in their capacity as musicians to brighten up camp life. Fortunately, they were fully equipped with instruments and they knew Glenn Miller, who was then at his peak; so much so that I am still a fan. With the 'happy garden' (Japanese jargon) now filled to capacity — over 1,000 — a new atmosphere prevailed. Even when working well, essential services in the camp were inadequate and antiquated. For the entire camp less than twenty WCs and about twenty-five very small wash basins were available. The kitchen was badly equipped, although only required to handle two meals each day at eleven am and five thirty pm. It is far from pleasant to contemplate what conditions will be like with the approach of summer.

On the lighter side, a big effort is being made to cater for those musically inclined. There are over twenty good musicians and an able conductor. Most of the music will be swing and from practice it would appear the band has some 'hot' numbers in its repertoire. They have been putting in some practice also for Sunday Church Service. Much amusement was caused by their first attempt at the doxology. The tendency to swing it was very noticeable. As for the guitars, they were very peeved when asked to drop out. To get the full significance of our band in top form one has to recall the attitude of an American jazz band with its maestro.

On the credit side, hot water showers have just been introduced. Let it be said, to begin with at any rate, there is a plentiful supply of hot water. With the prevailing weather it is possible to do physical exercises out of doors. The American instructor, who was rather unorthodox, caused much amusement, but nevertheless contributed considerably to the physical jerks. I feel the benefit of these exercises which, coupled with the simple life we lead, should keep one reasonably fit. There is a need to plan the day's activities. I managed to put in some work at Chinese character writing today. The education committee are working on an arrangement of subjects for which we are fortunate to have specialists in our midst. There is a definite improvement in camp cleanliness. The rooms look much neater despite the absence of shelves and facilities for storing.

Most important of all, the lavatories are kept clean and certain elementary rules are now well observed.

The religious side of camp life should be well taken care of now that we have twenty ordained ministers with us. An Anglican Communion Service is held at seven am and an Evangelical Service at three pm every Sunday. About 200 attend the latter, which is well supported by the musicians. The preacher this day was an American who took as his text 'Be of Good Cheer, I have overcome the world'. In a place such as this the Sunday Service ought to grow, both in the good it does and in attendance.

This is haircutting day, a highly commercialized business in certain sections of the camp. As much as $5 (Chinese) is being charged per session or $100 for the duration. Buying and selling is being 'pushed' in many ways, particularly articles of clothing and goods bought in the canteen. For the most part the money is used to cover gambling debts. An educational syllabus has now been issued and is very ambitious. Altogether there are about sixty courses, mostly in languages. It should be a stimulus to camp life and prove of value if we ever get back to normal. The whole scheme is in the capable hands of the professors from St John's University.

The weather continues to remain perfect. These are wonderful days if only we had our freedom and were able to get into the country. Lovely thoughts only. Our immediate prospect is to clear a large open space, known — in Japanese terminology — as the 'happy garden' but in reality a heap of bricks and a mass of old house foundations. This is our future recreation ground. Classes opened today to help the students keep in touch and to encourage the older ones to increase their knowledge. I enrolled for Chinese language, accountancy, shorthand, physiology and theology. Today a considerable area of the yard was covered with ordure. It seems to come up all over the place and those who had looked into the position say that the volume of the septic tank is much too small and cannot cope with the amount of soil where 1,000 people are involved.

A suicide greatly upset the camp today. A young American married to a Chinese was taken away from his wife and family. Prior to separation he was promised that keeping touch with them would not be a problem, but when 'cornered' here the Japanese would not allow communication of any kind. Sadly, this led to him taking his own life. A fine feature of the camp life was the attendance at Sunday Services. Many able men from a number of denominations carried the responsibility, in turn, of taking the Sunday Church Service. All are together, regardless of denomination, so that the substance of their message will have something in it for all who come to worship. This is a fine thing from which the camp should benefit.

The education business is going strong and seems to be doing well with practically no facilities. New classes are still being organized; the number of subjects for study must now be well over sixty. Today I attended a Bible Study course which promises to be very interesting. The physical features of Palestine together with its geographical position and the parts these play in the development of the Hebrew Race were outlined in class today. In the afternoon I made my first acquaintance with shorthand and book-keeping. Languages are very popular, particularly Russian, Spanish and French. On the unusual side of languages, Dutch, Malay, ancient and modern Greek are on the syllabus. Public speaking has not been left out and, believe it or not, modern efficiency methods. The camp in itself could well be a study in methods as well as personalities in the many and varied jobs encountered daily. A noticeable feature among the campers are the willing hands, practical and impractical and, in a category all by themselves, the bone-lazy. Truly this place is a study that would repay psychological attention.

Chinese employees have completed the work of making a connection to carry surplus ordure from our septic tank to a surface drain. Only in a state of war could such a shoddy job be undertaken.

The Saturday evening concert has come to stay. The band improves with practice, but some of the turns leave a little to be desired. However, under the circumstances — full marks! An entertainment lasting one and a half hours does a world of good. The musicians evidently enjoy themselves, especially the Negroes. Lights out was an hour later, which must have been special for the guards too. Gregg's shorthand is now part of my education. It is amazing how eight different classes can be taught with over 100 pupils all in the same room with no books or appliances to aid the teachers. The education committee are to be congratulated.

I no longer serve food in the dining room but am engaged

on the much more agreeable job of maintenance work, first of all on the roof, making an inspection with a view to keeping the weather off those below. Being in the practical category now I feel more useful.

More new internees are due today, altogether 112, which makes the total number well over 1,000. While they are welcome as harbingers with news about the conflict in the Pacific and Europe, our immediate concern is one of space in the camp. Morning classes have been cancelled by the authority in charge as labour for essential fatigue work is insufficient to keep the camp running smoothly. This is regrettable as undoubtedly the decision will make it impossible to arrange all the subjects so well planned by the committee.

The church service this Sunday was well attended and the text taken from Micah in the old testament was unfolded in the manner and depth of feeling worthy of it:

'What doth the Lord require of thee but to do justly, to love mercy and to walk humbly with Thy God?'

It was said by one of us, who had a previous connection with the block of buildings now regarded as our home, that they were condemned for the storage of tobacco ten years ago.

My letter to Rena in Australia, to be transmitted by the Red Cross, which had been in the hands of the camp authority for three weeks, has been returned with the whole completely censored. It is difficult to understand what one could write about and have accepted. The food not only in variety but in bulk is gradually going 'down-hill', so much so that a protest in writing has been made to the commandant! We were 'without rice' for several days and had to exist on a handful of inferior vegetables twice a day with a very small amount of horsemeat. Just when rations were at an all-time low, and feelings likewise, the Japanese aggravated our plight by announcing their intention to remove two Britons from our camp to, as they said, 'another place'. Where, we were left to guess, but no doubt it was for discipline. We were soon to observe the men in question being escorted from the camp under heavy guard. This led to a demonstration of cheers in the first place, then singing of 'There will always be an England'. As a consequence, the guard was doubled and the rooms were under continuous inspection, night and day. The camp commandant was dismissed and his successor made it known at once that he would take extreme measures if trouble arose. However, the food ration increased thereafter and rice was made available daily. A very interesting departure concerned those with Asiatic wives. Arrangements were made to enable communication, with a possibility of seeing them. For those with families far away the International Red Cross would come into the picture with mail and parcels. We must wait and see how long this latter concession takes to materialize. The first parcels from the International Red Cross arrived at the beginning of the month. Shell were associated with separate parcels for their staff.

Rumours are circulating thick and fast about the sailing of the *Conte Verde*, an Italian liner anchored in the Whangpu, which can be seen from the camp, receiving a coat of paint. Some time later repatriation was arranged for Americans, only on an exchange basis. One hundred and fifty left our camp on a ship bound for Goa in India.

A considerable amount of gambling helped to pass the time for some. Money and goods changed hands at crap and poker. Canned goods and personal effects are sold and raffled to make good the debts. Camp conditions during the last few days received two shocks. Over 10,000 lbs of cracked wheat arrived, making it possible to serve porridge at every breakfast for some time to come. On the other hand, two of the lads were caught conversing over the fence with a Chinese female. The trick employed to catch them was rather below the belt. Uniformed guards clad in Chinese clothes loitered among the Chinese outside the boundary. The lads fell into the trap and as a punishment were forced to kneel in the commandant's office for the rest of the day. It was to be a three day kneeling punishment without food, but for the intervention of the British representative. The whole camp was punished by closing the exercise ground for two days.

Easter has come and gone but it brought a pleasant break in the otherwise dull life of the camp. A religious service on Good Friday took place in the afternoon. On Sunday there were a number of services. The Roman Catholics have now come into their own with the arrival of two priests to look after their spiritual needs. It is said that the two fellows just mentioned entered the camp voluntarily. The Sunday afternoon service was well attended by about 400. Both the singing and the address reflected the Easter message, which was well received.

Information from other camps in the vicinity of Shanghai reveals considerably better conditions there than with us. The drab warehouse atmosphere and surroundings of Pootung are a contrast to Lungwha, which is more in the country.

The arrival of the second batch of parcels three months after internment was quite an event. This was followed by a visit from the IRC representative who informed us more parcels may be expected and that 30,000 were actually awaiting delivery to all camps. The contents of the parcels were well chosen and most acceptable. The weather is now very warm and shorts are the order of the day. Sunshine is already taking its toll and by the end of the season, with shorts our only attire, the place will look as if south-sea islanders have taken over.

A lot of hard work has gone into a large section of the recreation ground, so much so that the playing fields are now quite flat and ready for the purpose for which they were originally intended — to play ball! Beyond, gardens have been laid out for those with the energy and know-how to grow vegetables. My weight is now 159 lbs. This is gratifying but has only been made possible through having outside food parcels, not the camp diet. Exercise is needed. Thank goodness for a new PT class in preparation for taking the field at soccer, in centre forward position, to celebrate the opening of the recreation ground. My second game at football proved disastrous. Captaining the team in a nine-a-side game, I went all out and while able to stay the pace, a sharp knock on the muscle above the knee, near the end of the game, rendered me *hors de combat*. In fact, the pain as a result spread to the ankle. This put an end to football for me.

A number of questions were prepared in readiness for a visit from the Swiss consul. Regarding the all-important matter of correspondence, the consul promised help with the local authorities in connection with quicker despatch and more frequent opportunities for writing. In six months we have only been allowed to write to our families once. Getting letters to foreign countries at war with Japan was understandably difficult, although it seems to me the International Red Cross have a part to play here. The sad part of this story concerns wives and children of neutrals who were completely cut off only a few miles distant. Our captors were adamant in the application of their no-communication policy. Occasion was taken from time to time, when Chinese workmen were in the camp, to smuggle letters out. When discovered the Chinese, doing a good turn, were horribly tortured for helping husbands to communicate news of their welfare to anxious wives not far away.

The educational committee is running a summer school. The school has been a great boon, in fact the salvation of the camp's majority. The awful monotony of the life tended to drag down many of us, mentally as well as physically. In a large measure this has been kept at bay by the effort of the splendid teaching talent among us to make their skills available on a great variety of subjects. Apart from classes, a fine series of talks has been running every second weekend. The titles reveal the gifted lecturers: food nutrition; a trip along the Yangtse; islands of the South Seas, history, including natural history; finger prints in criminal investigation; Confucionism, to mention but a few. An art exhibition is planned.

No account on the educational-cum-leisure picture would be complete without mention of music every second Saturday evening — the band from the *President* liner have truly carved a niche for themselves in the life of the camp. At swing or classics, their show is great and appreciated. The latest find, in the shape of a regimental band conductor, laid the foundation for a really good semi-classical concert in which Gilbert and Sullivan, *Il Trovatore* and Schubert's Unfinished Symphony played a part.

Church services have had many prominent speakers addressing an average attendance of 150-200. These services are an essential part and fill a real need in the camp life. Today, 6 July, was a red letter one for me. I had eight letters through the prisoners of war post.

'As water is to a thirsty traveller in the desert so is news from a far country to one who has been longing for it for ages'. The letters are from a year to nine months old, but how very welcome!

Towards the end of July and up to the present, the thermometer has rarely been over 80°F. This is most unusual for summer in Shanghai. Rain has played a big part in keeping us cool. One afternoon the clouds opened and the resultant downpour revealed the frailty of our tobacco warehouse in all its nakedness. Rain poured in through roofs, windows and dilapidated brickwork. Our chief anxiety was to protect a very precious electric wall-plug. This under-the-punka-creation of power was a great boon in that it was the source which enabled a number of us, within a certain orbit of the point, to get hot-water, cook eggs any style and bake a bread pudding. Our bread, lacking an important ingredient, was unable to stay edible longer than a few hours, hence the need for this vital organ which in our opinion was priority number one for protection from the rain.

A few days ago I received another letter, through the Red Cross, from Australia. It was fine to know that Rena and the wee ones were well.

The strange disappearance of an American from our midst was very mysterious. Admittedly he was regarded, by some of us, as having more than a little Japanese in his make-up. Nevertheless, as one of us, he was suddenly taken away and has not been heard of since. The second unexplained incident concerned a Briton who 'vanished' for twenty-four hours. It is said that he parted with a large haul of gold dollars which perhaps could be more accurately described as 'loot'. Anyway, it was to 'save his skin' so he is now one of us again. About this time too the disappearance of another kind is worthy of note. All of the foregoing in this paragraph gives an idea of the small world we were living in compared with the major events taking place daily in the big world outside. But to conclude with the last of the trivialities, our sole contact outside of the camp was *The Shanghai Times*, a fearful 'rag' full of propaganda of the deepest dye! After months of daily appearance, circulation of this paper stopped without explanation.

American repatriation hopes have risen to a new high. The Japanese officially announced an exchange has been arranged involving 1,500 US nationals from East Asia. In numbers, as far as Pootung Camp is concerned, this means 150. British hopes of repatriation have risen slightly with an announcement that negotiations are in progress. Most of us feel, however, that our chances are very slim. The departing Americans were allowed ninety kilos of baggage. Printed matter in any shape or form was absolutely forbidden. A piece by piece examination was carried out and many articles were removed. The party leaves for the ship on Sunday 19 September 1943. A farewell concert was a great success, with goodbye items from the choir and band. It was a splendid example of British/American cooperation. Many of the choir members were American. About eighty per cent of the teaching fraternity of our 'University' are on the way home. This will be a great loss, though a big effort is being made to keep the classes going.

Long before daylight on the day of departure we were astir. Indeed, just as daylight was breaking a Scot playing the bagpipes marched around the building giving them a send-off with an appropriate farewell tune. The 150 en route to the ship assembled for their last 'grilling' before departure by consular police, Japanese customs' officials and gendarmerie. By eleven thirty am, to the tune of a Hawaiian Farewell, the boys 'pushed off'. How many of us would like to have been going with them we scarcely dared say.

Three camps at Yangchow (Yangzhou), just north of Chingkiang (Zhenjiang) on the River Yangtse, are to be closed due to military necessity. One camp is transferred to Weihsien (Weixian) in Shantung (Shandong), with the North China contingent and the other two to be with us. The prospect of having women in the camp fell on us like a bomb-shell. In the first place, this is a horrible location for any woman. Secondly, in no way can privacy for families be envisaged. Further complete alterations will have to be made for lavatory and washing accommodation. Time will tell, but right now it is difficult to picture such a change for the 405 newcomers. To make room for this number, the opposite sex predominating, a transfer of 250 men from Pootung to other camps was necessary. This number was named by the Japanese commandant here from troublemakers. This method of selection was quite acceptable to those who remained. The Yangchow (Yangzhow) contingent arrived after a distressing journey by launch, bus and train and three nights on the floor with no bedding. The big problem, still unsolved, is lavatory facilities. Now, with women in the camp, the situation in this connection is disgraceful. To say the least of it, the whole sanitary system is completely inadequate. The effluence frequently chokes, with disastrous results. Enough said; the situation has to be seen to be believed.

These are lovely autumn days and life out of doors is pleasant. Oh to be out of this place! The meagre news that filters in from outside suggests that a silver lining is showing up on the international front, which is very heartening to us here.

Medical attention within the camp was in the capable hands of two missionary doctors. They had a clinic and a sick-bay with nursing assistance for anyone who needed special care. The Japanese were very reluctant to give their permission for outside hospital care, even if pressed by camp doctors. Acute colitis has been my lot in the last ten days, a devastating sickness which pulled me down no end. Roughage in the food was much in excess of what our tender stomachs could cope with. For me then, the roughage was more than I could take, with the result it ripped the lining off the intestine, causing much pain, loss of blood and inability to eat. The doctors were able to persuade the powers that be I needed outside hospital care. On their part they thought it unlikely I would return, so on a stretcher I left the camp for the General Hospital in Shanghai. What a joy it was to get into hospital. The nuns, as nurses, made me feel it was more like heaven.

I left Pootung internment camp for hospital 10 October 1943 and it was nearing Christmas before I began to show any signs of improvement. In the meantime, it was possible for me to contact friends outside — the Zings, already mentioned, and my old Chinese language teacher, Miao. It was really wonderful to see them and how good they were to me in money and kind. Miao was especially helpful. I wrote many letters to Rena and the wee ones in Australia. I need hardly say that under no circumstances could mail be sent 'down-under' from Shanghai. Just how these letters found their way to Australia is a mystery to me. But I know they travelled many miles westward before reaching unoccupied China, all free of charge to the writer. How do you pay such a debt?

As Christmas approached the Japanese, who were advised from time to time on the progress of prisoners in hospital, decided my return to their jail was overdue. One day the doctor, who was a Czech, stopped by my bedside to say he was about to agree on my return. Then as an aside he whispered, 'Do you still have your tonsils or appendix?' My reply was, 'I have both.' The next one took my breath away. 'Well,' said the medico, 'which of these would you like removed so that I can keep you in hospital for Christmas?' I reacted immediately to this wonderful suggestion with, 'Tonsils, please!' Little did I know at the advanced age of forty tonsils, still in place, can be very tough. I would not have agreed to the doctor's proposal and when all was ready he handed me a looking-glass with the remark, 'Would you like to see the operation?' In a blasé mood

I decided to 'play the game', much to my later regret. Half way through, when the blood was all over the place, I fainted. Was the doctor mad? He had to revive the patient before proceeding further. Needless to say, without the use of a mirror on my part, when all was said and done — it was worth it to have Christmas in the hospital. Soon afterwards, I was back again in the rough and tumble.

From now on incarceration Japanese-style in Pootung had an effect on me which played an important part in my life even after internment. On return from hospital I was asked to work in the clinic participating in vaccinations, inoculations and generally assisting the two doctors in their work of keeping sickness at bay with over 1,000 on their 'panel' who were poorly housed and badly fed. I attended a course in the principles and practice of first aid to the syllabus of St John's Ambulance Brigade and have their certificate. Occasion did arise when the whole camp had to be injected with a hypodermic syringe when a sickness was in the vicinity and all were in need of protection. At first finding the vein with the needle was a minor problem.

One day a surprising proposal came my way, out of the blue so to speak. The camp had been practically starved of dental attention. The occasional visit of a Japanese dentist did little more than scratch the surface. On such a poor diet the condition of the campers' teeth can well be imagined. The doctors had no dental experience — at any rate they were fully extended taking care of other diseases. But to get down to the nitty-gritty, early in 1944 one of the doctors approached me with a proposal to set up a dental-clinic. The Japanese 'dumped' the equipment on our 'lap', which they had 'acquired' from some source in Shanghai. At this point I asked the doctor what made him think of me. His reply was that there is a lot of engineering in dentistry — drilling, filling and extraction etc. My reply was that surely I must know something of the anatomy in the area of the mouth and, to put it mildly, how far could I go? In a final effort to try and get out of it I pointed to the lack of anaesthetics, to which the doctor replied, 'Switzerland, the UK protecting power, will supply us'. On the understanding that the doctor would do the extractions and come to my aid if something arose that was beyond me, I accepted the job. No need to worry now about passing the time away. After all, 1,000 patients on the doorstep.

As I steer my way through the many and varied problems

The Prison Camp dentist in action

associated with dentistry and the use made of this clinic for another purpose, I trust readers will forgive the use of gory detail from time to time. I was but a novice in the clinic when both doctor and a patient, sitting in the chair, faced me with no alternative but to show that I could do without hesitation and indeed with confidence, what was required. The patient was a sailor from a captured American ship and he was suffering from the hell of all diseases with little or nothing left above the gum of the offending molar. After 'boiling' the tools and preparing the syringe with the necessary ccs. of anaesthetic I called the doctor. Imagine my surprise when, in front of the patient, he looked at me and said, 'Go ahead.' It was a never to be forgotten moment in my life. To say 'No!' or even hesitate would have been fatal. The patient was watching and he would undoubtedly have informed the camp, 'Beware of this "quack" ' had I shown the slightest sign of passing the 'buck' back to the doctor. Using a disinfectant, I swabbed the gum in way of the molar then two ccs of novocaine with the syringe. During the waiting time for the painkiller to take effect I had to work out just how it would be practical to get a firm grip of the offending molar using the extracting forceps, bearing in mind little or nothing could be seen of the tooth above the gum. When the patient indicated only a numbness in way of the extraction I picked up a scalpel and slit the gum on both sides of the offending tooth, which enable me to grip it with the forceps down to the root. Then with my head over his shoulders and eyes closed, I prayed for help by way of a strong wrist motion. Suffice it to say that this uprooting was the first of many, but for now let us change the subject.

In the first days after 7 December 1941 it will be recalled I was living in a flat, over Shell's head office in Shanghai. My neighbour and close colleague was a radio 'ham' whose enthusiasm knew no bounds, even to the extent of allowing his Chinese girl friend to send radio messages which could only be interpreted by the Japanese as espionage. Now we were in the same lock-up and he often regretted not having the equipment to make a radio. We were really starved for news and it would have been a godsend, not only to relieve the monotony but to boost the morale, to be in radio communication. My work in the dental clinic involved a very occasional visit to Shanghai, accompanied by a Japanese guard, to purchase dental equipment and contact the Swiss for anaesthetic. The guard took the opportunity to see his friends, and left me at an address with Chinese friends from where he picked me up later. This happened several times, which prompted discussions with my radio colleague when we worked out the equipment needed, valves, wiring etc., to build a radio receiver capable of picking up San Francisco. While in Shanghai with the guard elsewhere, it was left to me to work out ways and means of procuring material for the job. When all was obtained and smuggled into the camp, the next problem was where and at what time of day could we work in secrecy to achieve the very much desired result.

The Japanese guards were all over the place at unexpected times. For obvious reasons, a quiet place with no need to cover up quickly was essential. It seemed to me there was nothing better than the dental clinic, after hours when the camp activity was very much alive and end of the day guard activity was at a low ebb. The brains behind the assembly was my colleague and we had to keep our entry and exit from the clinic private to the nth degree. The radio parts were safely stowed in the clinic during working hours so we were all set to go when the area was quiet. The dental drilling machine and drills for removing decay proved absolutely necessary in our assembly work. We had very much in mind the need to make this receiver not only suitable for the job but compact for storage, in a place still to be found, when not in use. Secrecy was maintained and the vital moment when we heard San Francisco calling, never to be forgotten. Suitable out of sight storage for the precious set was found near my bunk after removing a number of bricks from the wall and making a suitable cavity.

The dissemination of news from the source now in our homes was in need of careful consideration. It is no exaggeration to say we were handling dynamite. Shell colleagues, without hesitation, would share our secret. There were some twenty of us in Pootung and then, with advice from our associates, others were admitted to the circle. No doubt there would be a slip beyond this orbit but our aim was secrecy as much as possible, and this was not easy under the circumstances. However, the radio was with us until the atom bomb fell on Hiroshima; afterwards there was no need for it.

Concluding my assay into the field of dentistry, I still have in mind the agony of patient and dentist when the latter tried to extract an abscessed tooth — never again! On a certain examination carried out initially by flattening the tongue, I could not see the uvula above the root of the tongue. The absence of something that ought to be there was beyond me so I sent out a discreet SOS for the doctor. After swabbing what used to be the place of the uvula with gentian violet, we arranged to see the patient later. I could not question the doctor quick enough and he said the absence of the uvula could be explained by either 'trench mouth' or cancer. My patient died three months later. Another case involved the aftermath of an attempted escape. A young man in his late teens had enough of confinement and made up his mind to go for freedom. A very long walk lay ahead of him, although, had he made it, I am sure the Chinese he met en route would have been helpful. As it was, he did not get beyond the tight ring the Japanese maintained around Shanghai. On return to the camp the guards who had lost a lot of face in not preventing the escape used a baseball bat to beat him up. One of the blows struck him on the mouth and broke many of his teeth. My job was to make life more comfortable for him. I won't go into the details but it was very painful for both of us. Shell colleagues of mine swore they would rather die than have me attend their teeth. Time was on my side! Finally, I have a super satirical cartoon, drawn by a patient, of a certain amateur dentist performing on an irate subject roped in place for the operation. The picture graces the bathroom in our present home. The medical officer in the camp was kind enough to write a letter and a report on my 'performance' as a dentist. Both are part of this story.

During 1944 American Red Cross parcels found their way to Pootung camp. Here is a quotation from my diary about this 'manna from heaven'. Coming about Easter time, between Good Friday and Easter Sunday, it was a rare gift. Everything contained eatables of a very high food value. Then follows a list of articles. I recall one item headed 'Emergency Ration; ingredients: chocolate, sugar, skim-milk powder, cocoa fat, oat flour, vitamin B.

4 ounces net — 600 calories

To be eaten slowly (in about half an hour).'

With better news of what was going on in the world outside, thanks to our radio, the passage of time brought with it a degree of impatience. A thought never far from our minds was: will the Japanese leave us here or will they break up the camps and send some of us to Japan? From time to time in 1945 we saw and heard American planes bombing Shanghai. Our morale was on the up and up, but doubts as to our ultimate fate persisted. Much has been said about the horrific effect of the atom bomb on Hiroshima on 6 August 1945 and again on Nagasaki, but in reality there is no doubt that as a result many lives were saved, especially in North China.

Had I been asked to put into as few words as possible on how I would like this nightmare to end it would have been with thanks to God and this thought uppermost in mind — oh for one morning to wake up and find our guards had disappeared overnight. This actually happened on a certain morning in late August 1945.

Thereafter, my thoughts were very much concerned with when I could get home. It was five years since last I saw my wife and the wee ones. Little did I know, when leaving camp, that they had left Australia when the war in Europe was over in the Spring of 1945. With me, Shell obviously wanted to get their business started in China even if incoming oil shipments were a long way off. For those in a hurry to get home, transportation was a problem. It is a long way from East Asia to these islands in the west of Europe. So office work it had to be, even if just to show willing. It was some time in October before it was possible to get a passage, and then on an aircraft carrier from Hong Kong with below deck accommodation, sleeping in a hammock. It was a stormy passage and we were not sure when and where we would arrive. Indeed, the officer in charge of the passengers made it quite clear, after we had been advised that Liverpool was the port of entry, that it was not convenient for anyone to meet us at the dock. Accordingly Glasgow Central Station was the chosen spot for meeting with Rena and Mother, the latter whom I had not seen for ten years.

7

Following Japanese Defeat in Asia and the Pacific

Having accounted for what happened to me, in some detail, after war was declared in the Pacific, what was the lot of my family who were in Australia for four years? The Japanese midget submarine which showed up and fired shots in Sydney harbour shook everyone to the core, so much so that many of the 'dinkum Aussies' left for the mountains. This evacuation made it possible to rent a suitable bungalow in the Rose Bay area. Prior to this they had been living in a flat at Vaucluse near the South Head of the harbour. Almost without exception the neighbours were friendly and helpful. Rena assisted in a Servicemen's Club while the children made friends at school and laid the foundation for their early education, with careful emphasis from Mother on the speaking of English without an Australian accent. On one occasion they had the frightening experience of a late night intruder but a neighbour came to the rescue. The garden needed a lot of attention. This helped to pass the time and Shell wives were always at hand to keep in touch.

One memorable day a good number of letters arrived from Father, with a Chungking (Chongqing) postmark, written during the time I was in hospital. The despatch was arranged by my old Chinese teacher who used a messenger system to bypass Chinese territory occupied by the Japanese. The family were faithful in their church attendance, almost every Sunday. Apart from the minister they were seldom spoken to, which says very little for the Methodists of Sydney, at least in two quite separate congregations.

When the war was over in Europe an opportunity arose through Shell for Rena and the girls to get a passage home. Much of the long sea voyage via the Panama Canal was in the Pacific, still a war zone. It was a crowded ship with six in a cabin. But for the fact they were on the way home, it was more of an ordeal than a pleasant journey. Those at home were unaware that the family from Australia were on their way back to Blighty. Imagine the surprise when Grandpa Yuille opened his front door one morning at Beechwood Terrace, Dumbarton and, there standing in front of him were his daughter and grandchildren from Australia. His letters had been a great strength to Rena in the dark days when the Japanese were advancing in the Pacific. He always likened them to an octopus whose tentacles withered on stretching too far. As is well known, this is what happened to the Nipponese.

After such a long period of absence, most of my time at home was spent renewing acquaintance with family and relations: my brother Willie, a schoolmaster at Glen Luce, Wigtownshire, and sister Elise, married to a Church of Scotland Minister and living at New Deer, Aberdeenshire. We all stayed at the manse for a short period and enjoyed the quiet and meeting folks whose whole way of living and outlook was so different from ours. While there I had a call from Shell to go to London and will always remember an encounter at Aberdeen railway station as I was trying to get a sleeper on the train south. A girl was handling the ticket office responsibility and she kept saying in her best high school English. 'You cannot get a sleeper.' The need for such a berth over a 500 mile journey made me push with a little more pressure of words than I normally use. Finally she said, for all in the vicinity to here, 'A canna gie ye whit a hane got'. With that I could do no other than sit up all the way.

My Mother at seventy years of age was hale and hearty in all her activities. The church and happenings during the war under its auspices was oft repeated. Her home baking on Saturday morning was an activity still very much part and parcel of her life and was much appreciated by all who partook of the result. My enthusiasm for having the radio always turned on was not to her liking and it was frequently silenced. Mother always let it be known she was a 'Liberal' in politics and she once allowed her name to go forward as proposer of the 'Liberal' candidate for the town council election. Relatives thoroughly disapproved and she had to withdraw. Father, on the other hand, never let it be known how he stood on politics. In fact, he rarely discussed the subject.

I returned to China by air in the spring of 1946, all the way by the well-seasoned and thoroughly reliable Dakota! Having taken the punishment of the war years, this aircraft was used worldwide by civil aviation to bridge the gap awaiting the comfort and speed of the jet-engined plane. It was the season of the SW monsoon in the Indian Ocean, my first experience of flying, and it was awesome to observe from the plane that it was only at an elevation of about 100 feet above the waves from Calcutta to Rangoon. It was a relief to get to Hong Kong with only a few hours more in the air to my final destination, Shanghai.

Rena and our two young daughters were left in Dumbarton for another spell of separation. According to Shell the situation in North China was not yet stable enough for their return. If the truth were only known, Shell had in mind for me a protracted trip of 1,500 miles along the Yangtse Valley to make detailed inventory of tankage, buildings and equipment at their installations.

This was a major job and with the approach of summer, travelling light was the order of the day. The job involved little or no social round apart from essential contact with former Chinese staff. It should be borne in mind that for a period of six years our installations were unstaffed and in consequence open to all kinds of malpractice. But to cut a long story short, I will deal only with the highlights. In the smaller ports, Chingkiang (Zhenjiang), Wuhu and Kiukiang (Jiujiang), the plants, including tankage, were in need of only maintenance and spare parts. Indeed it was surprising how unattended property had been respected. It was only on arrival at the major distribution port of Hankow (Wuhan) that a surprise awaited me.

On the north bank of the Yangtse (Changjiang) about three miles down-stream from Hankow (Wuhan), Shell Oil and American competition, Standard Oil and Texaco, had their installations within a stone's throw of each other. On my arrival at the Shell site I was greeted with a sight that left me speechless. The plant I had come to inspect was partially destroyed by a bombing operation from the air. This was confirmed by former Chinese staff. The astonishing thing was that the two other opposition installations were obviously untouched when bombs had been dropped on Shell. A conclusion was beyond me, but I did mentally register that should I come across a similar happening that it must be dealt with officially.

Ichang (Yichang) was the furthest point west in Japanese-occupied China. Apart from depots, this was my next major plant to be inspected. The position of the three oil companies' installations was cheek by jowl, as in Hankow (Wuhan) and a bombing raid resulted in the same devastation for Shell. It is inconceivable to associate the Japanese with such discrimination. Had they decided to do such a thing it would have been done as a whole. In consequence, I could come to no other conclusion than that it was planned by the high command in Chungking (Chongqing). I never did put this very delicate matter in writing but on reporting verbally Shell vetoed any further comments.

May 1946

On completion of the survey work on the oil installation at Ichang (Yichang), 1,000 miles along the Yangtse river from the coast, the next stage of travel was 340 miles through the gorges and rapids of the Upper Yangtse to Chungking (Chongqing). It was not my first experience on the upper river, but this time I was in for a shock. Twelve years before, the Black family made a similar journey but that story has already been told. Fortunately, so I thought at the time, I received an invitation from the captain of a British gunboat to join him, as a passenger, on board his ship bound for the same destination. My curiosity was aroused when I thought, how is it possible for a ship flying the White Ensign to be in these waters? The captain explained: the British Ambassador, in need of prestige from the Chinese Government in Nanking (Nanking), offered to put a gunboat at their disposal to provide safe passage for records from their wartime capital Chungking (Chongqing).

One day early in May 1946 we sailed from Ichang (Yichang) with two non-English-speaking Chinese pilots. However, as their instructions to men at the steering wheel was by finger motion, speech was unnecessary. With experience in this section of the river over a number of years, I was well aware that this type of vessel was not of the design used in the days of 'gunboat diplomacy' on the Upper Yangtse. Information on leaving did not indicate any sign of a rapid rise in the upper reaches of the river. It was only when we approached the Hsin T'an rapid, some thirty-four miles from Ichang (Yichang) that the strength of the current was in excess of ten knots. Did this vessel have the power to make headway against such a current? As rocks were strewn on the banks on either side of the river, it was no exaggeration to say that a dangerous situation existed.

As I paced the main deck, below the navigating bridge, looking out on the scene just described, I was very conscious that the ship and all on board were in a precarious position. We were making little or no headway on full power but what frightened me was the weaving of the ship from one river bank to the other. Just as it was about to strike the rocks the ship would answer to steering and plunge across the river to the opposite bank. What a relief! But would it continue to escape what seemed to me the inevitable? At this point a sailor came from the bridge with a message from the Captain that he would like to see me. On the bridge the look on all faces, even the pilots, normally inscrutable, showed signs of considerable unease. I was in no position to provide any assistance, from the navigational point of view, but the captain asked if I would speak to the pilots and get their reaction. They immediately responded, I well remember the words in Chinese: it is very dangerous, the ship must return to Ichang (Yichang). As can well be imagined, it was impossible to swing the ship. The method employed was fraught with danger — a slow astern, with trust in the steering, and let the current take over.

We never did get to Chungking (Chongqing) and the Chinese Government had to wait a little longer for their records. In actual fact, the vessel sent by the Admiralty was not suitable; she had the power but the number of rudders and the steering area on each rudder was not designed for shallow draft and strong currents. I had to return, nearly 700 miles by river to Hankow (Wuhan) and take a plane for Chungking (Chongqing), but that is another story.

The journey by air to my final destination, Chungking (Chongqing), from Hankow (Wuhan) was by a plane ill-equipped for passengers. There were no seats and consequently no belts. All luggage was carried with us and we squatted on the floor. Given a smooth passage, this far from comfortable way of travel by air would not have been too bad, but it turned out that the going was rough and without seat belts passengers and luggage were all over the place. It is hard to picture such a situation in these days of air travel comfort and service.

Chungking at last! There are few or no problems here in this the capital of unoccupied China for nearly five years. Shell management in Shanghai asked me to entertain the Chinese staff of the Standard Oil Company, who gave Shell considerable assistance in fuelling the aircraft of our clients. Having sent out all the invitations to an all male Chinese lunch, I was somewhat taken aback by a request from their manager when he asked, 'May my wife accompany me?' Though puzzled, I could see no reason to object. Once we began eating and in accordance with Chinese custom, toasting every dish with a small wine cupful of 'Mao-tai', I could see the reason for the only lady at the party. She was in Chinese parlance a 'ti-kung' — a substitute for her husband when it came to drinking 'Mao Tai'. She seemed to delight in seeing her fellow countrymen below the table. Earlier on I made up my mind to see how she left the table and the restaurant at the end of the meal. This was accomplished with an astonishing panache. Outside a rickshaw was waiting and off she went, a triumph of the occasion.

1947 introduced the period destined to be the outstanding one of my all but twenty years with Shell in China. At the beginning of the year it was a matter of pursuing problems arising out of my long journey along the Yangtse Valley. The distribution of oil products by road, rail and water was far from normal in comparison with December 1941 when Japan soon after occupied China in depth. At about this time too Shell decided to discontinue the use of the old name, Asiatic Petroleum Company for North and South China, and in its place use Shell Petroleum, with head office in Shanghai. Staff problems were numerous, especially when transfers were made from semi-tropical Hong Kong to Beijing, formerly Peking and Tientsin (Tianjin), where the sand of the Gobi was very much part of life and winter temperatures were well below zero.

Even then the main oil product was still kerosene (paraffin), or oil for the lamps of China. I never ceased to be astonished at the lack of progress made by central government in the introduction of electricity for both power and light over the enormous hinterland of China. It is true to say that in the twenty year period between the world wars the central government was rarely in control of the whole country. War-lords ruled the roost in many parts and they levied taxes for their own use. The oil companies had an interest in selling kerosene and it could be that their influence was used to keep this product in the forefront. On the other hand, the need for heavier fuels to raise steam for turbines and drive diesel engines was apparent enough to make one wonder what was in the mind of central government. The Kuomintang party was again in power at Nanking (Nanjing) but in distant Manchuria the Communists were gathering strength and slowly moving south.

But to get back the point I made about 1947 being a milestone in my career with Shell, it was almost a year since I left home without Rena and the girls and I found it difficult to understand Shell's ruling on the subject that conditions in Shanghai were not yet settled enough for family life. Admittedly my long spell away on a survey of Yangtse Valley installations would have involved separation and it was better they should be at home. Now, however, as I saw the situation, it was time to have some home life together in China. Little did I know that Shell had another journey in mind for me — a journey to NW China in search of oil.

8

Journey to NW China, Tibet and Gobi Desert followed by brief visit to USA and The Netherlands

In the Spring of 1947 Dr Wang Wen Hao, Head of the Natural Resources Commission in the Kuomintang Government, invited the management of Shell, Standard-Vacuum and Caltex to a meeting with his own organization, the Chinese Petroleum Corporation, to discuss oil possibilities in NW China. His plan was for a team to take part in field work of some four months' duration including a traverse across the Chi lian shan (Zilian Shan) or Nan Shan. This mountain range with peaks of over 20,000 feet is the northern edge of the Tibetan Plateau whose foothills fade out into the Gobi Desert. It was also part of the original plan to carry out a month's exploration by yak caravan in the region of the Upper Tsaidam (Zaidam) Basin. It took some time for the oil companies concerned to get a team together of geologists and geophysicists. To my surprise Shell of China decided I should be one of the party, possibly because of language knowledge and experience of working in China. All four others were American and the team was headed by a geologist from Standard Vacuum.

While waiting for development and further planning of my Kansu (Gansu)/Chinghai (Zinghai) journey, I was off to Peking (Beijing) on company business. I stayed with a colleague. When time permitted my interest turned to renewing acquaintance with the palaces, pagodas and pavilions of this unique city's imperial past. Although it can be said that a city has grown up here for over 1,000 years, the finely attuned sense of colour and form it has today must be associated with Kublai Khan who mounted the Dragon Throne in 1269. For me it is one of the most fascinating places on earth. I left Peking (Beijing) with regret and it has to be said, as I write these lines now, never to return, although given the opportunity recently by a Chinese friend to be present at the opening of his new hotel in the capital.

Returning now, to what will frequently be referred to hereafter as the Kansu (Gansu) and Chinghai (Zinghai) Journey, the geologists and geophysicists arrived in Shanghai from America on 14 June 1947. At this point I joined the team in their meetings with principals and the National Resources Commission of the Chinese Government. It was early apparent that the plan outlined by Dr Wong Wen Hao which included much ground work and in consequence a considerable amount of time, did not meet with the experts' approval. They were all for aerial survey covering the traverse on the ground proposed by the NRC. In this connection photographic equipment had already been assembled in US and was part of their luggage. This proposal was accepted by the Chinese but the provision of a suitable aeroplane and the need to consider refuelling took some time to solve. This brought us into contact with General Chennault of China Air Transport Services. The Second World War General assured the party he could meet requirements by providing a C-47 plane with an experienced crew. In fact, he had a pilot available who had considerable experience 'flying the hump'. Could we have found anything more suitable for flying on the plateau of Tibet?

Over the period of July and August 1947 I kept a detailed diary covering the journey of the Joint Survey Party in the provinces of Kansu (Gansu), Chinghai (Zinghai) and Ningsia (Ningxia). While my diary of some 30,000 words is very much part of all that happened in this most interesting safari, in search of oil in Central Asia, I propose to deal only with the highlights here: in the first place where they concern the object of the party — the search for oil possibilities, and secondly the interesting people, places and things that crossed our path on the way.

The National Resources Commission of Nanking (Nanjing), sponsors of this trip, provided each foreign member of the party with permits to ensure safe conduct. Of special importance was there permit bearing the personal stamps of General Chiang Kai Shih in which he requested all authorities en route to facilitate the work of the party. Last but not least was an official document giving government permission for aerial photography. This included photographing the proposed pipeline trace from Hankow (Wuhan) to Lanchow (Lanzhou) in Kansu (Gansu) via Sian (Xian) and the main area of exploration, the northern escarpment of the Tibetan Plateau, then, on the plateau itself, the big lake Koko Nor (Zinghai Hu) and the Tsaidam Basin (Zaidam Pendi).

The accompanying map shows the route taken by plane on the first day of July 1947 to Lanchow (Lanzhou). Thereafter, it indicates the journey, giving the mileage, in the air and then by truck on the Old Silk Road and, most important of all, by camel and mule cart on caravan trails.

From now on north-west of Hankow (Wuhan), highlights from my diary are given. The valley of the Wei River on which Sian is situated was quite a revelation. This basin, well-known as the cradle of the Chinese race, is broad and obviously prosperous. Sian (Xian), the provincial capital of Shensi (Shenxi) enclosed within a wall, is in the centre of the valley. For a distance of 100 miles the fertile valley of the Wei was under observation from the plane. The southern Kansu (Gansu) plateau is 4,000 feet above sea level. Here the desolation of the loess country is a contrast to the cultivated land of the Wei valley. The Yellow River close at hand was a sign that Lanchow (Lanzhou) was near and by late afternoon the plane was circling the city. From the air, this city impresses one immediately, for the most part due to the peculiar behaviour of the Yellow River (Huang Ho). Looking as if it were a solid mass of mud moving along, the river breaks up and honeycombs the city. This, together with the mountains on all sides, makes it look very attractive, at any rate, from the air. The waterwheels and goatskin rafts have long been famed in story and travel accounts of this region. Irrigation is essential to life in such barren surroundings, the immense waterwheels, each unit scooping up hundreds of gallons a minute, making a large contribution to the welfare of the community. A newcomer cannot but be surprised at the sight of Arabic script over many shop windows until he recalls that the owners are likely to be Tung-kan merchants from nearby Sinkiang (Zinjiang) province.

Weather conditions were perfect when our C-47 took off from Lanchow (Lanzhou) at seven fifteen hours on 3 July for the flight

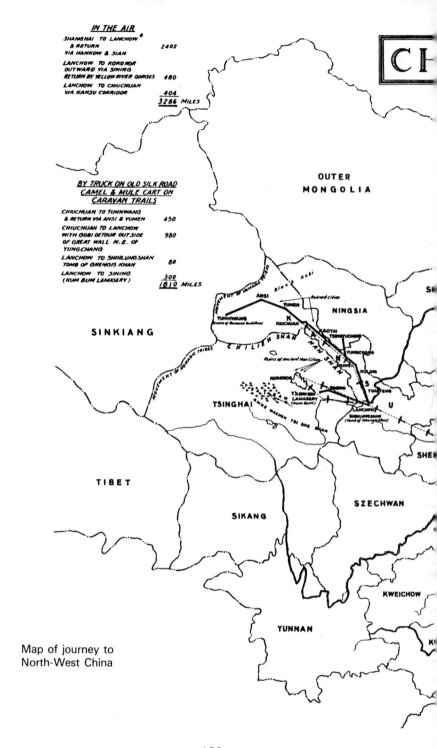

Map of journey to North-West China

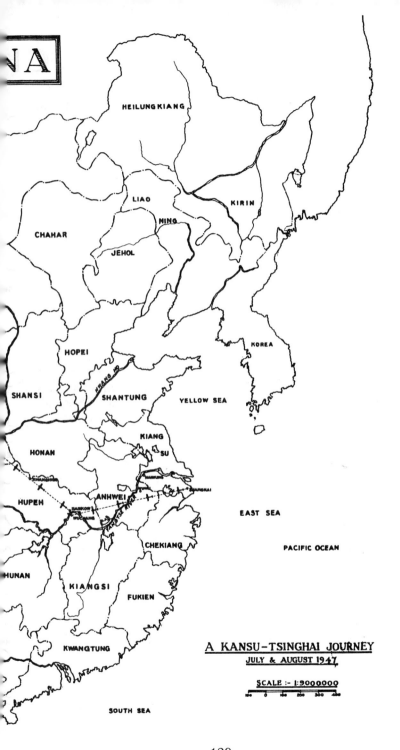

to the neighbouring province of Chinghai (Qinghai). The first leg of the flight lay along the valley of the Sining (Xining) River, a tributary of the Hwang (Huang) Ho. The strange sight of these great folds of desert mountains or loess can only be described as fantastic. The geologists of the party said they had never seen a panorama to compare with the extraordinary contours of this loess country. Windows of red sandstone could be seen at intervals along the river's edge. On the top of these red beds, centuries of sand storms from the Gobi had deposited layer upon layer of dust until it is now in places several hundred feet thick over the wastes of this desolate region, sometimes as high as 9,000 feet. Signs of cultivation were few and far between. The contrasting green stretches on the banks of the Sining (Xining) Ho, where wheat and barley grew, made even the term 'fantastic' seem out of place in this grotesque region.

The work on aerial pictures commenced here. As if by magic the loess mountain ranges came to an abrupt end and an almost perfect scene met the eye. The mountains, still high, were green to the summit and the loveliest valley, with a thread of river winding its way through green fields, lay below. Cultivation was everywhere, making nature's pattern of colours into a beautiful design. The sun was shining in the valley and the town of Sining (Xining), laid out in a square and enclosed with a wall, appeared as green as the valley and attractive.

Still following the Sining (Xining) Ho, by now just a stream near its source, the plane climbed to 11,000 feet. As we topped the rise and looked ahead, shining in the sunshine lay the Koko Nor (Qinghai Hu). From the cockpit of the plane the lake looked so vast and blue that a comparison with the Mediterranean was not out of place. For the second time today we were surprised beyond belief at the view which presented itself. Rolling grass lands stretched for mile after mile on the north west side of the lake, almost as far as the eye could see. As far as could be seen from the plane, these downlike uplands stretched to the horizon on the NW side of the lake. To the north a barrier of black cloud looked ominous and, being two hours out from our base, it was decided to return in case the cloudy weather should shut out Lanchow (Lanzhou). The plane then made a detour, turning back into the sunlight when cameras were in action and we again took note of the terrain. Hereabouts animals and birds were plentiful and the nomad beside his yurt, with a grazing herd under his eye, made a perfect picture on this last glimpse of the Tibetan Plateau. Of the many islands on the lake, the largest is Hai-sin-shan, with a length of 1,650 metres, breadth 560 metres and highest point 200 metres. Grass is abundant and there are more than ten temples under lama control. The lamas only leave their temple island when the lake is frozen. More than fifty streams flow into the lake. The Pu-ha-ho originating in the Nan Shan is about 100 miles in length. It should be mentioned for anyone who reads this and is 'afflicted' with an urge to explore, there is an area, practically undisturbed in the region of the lake, where three cities — Tsa-han, Tsiang-chun and Li-chen-tse, dating back to Han times (200 BC), await archaeological research.

The flight to Chiuchuan (Jiuquan) from Lanchow (Lanzhou) was the last leg of the outward trip, a distance of 500 miles. Practically the entire length of this section was of interest from a geological viewpoint so the journey was one of aerial photography. In the Chiuchuan (Jiuquan) — Laochunmiao (Laojunmiao) area the plane circled around the section of interest while intensive reconnaissance was made. The Nan Shan or Chi lian shan (Qilian) range of mountains was now between us and our earlier flight today, the Koko No on the plateau itself. Many of the peaks of this formidable range are permanently snow capped. In some places these giants rose to an elevation of 20,000 feet. The flight lay along the foothills of the Nan Shan, the last word means 'mountain', and followed the Old Silk Road along which coursed the trade of antiquity. To the north-east of this famous trail was a huge expanse of sand, the Gobi itself. The word Gobi is Mongolian in origin, it does not signify a certain geographical area, rather indicates a kind or type of desert.

In our flight along the northern escarpment of the Tibetan Plateau there was a lack of stability in the movement of the plane, indeed for almost the entire distance from Lanchow (Lanzhou) to Chiuchuan (Jiuquan) it was impossible to stand without a secure handgrip. Moisture laden clouds were not the explanation of these bumpy conditions, rather was it the result of violent thermal draughts caused by hot air rising from the desert meeting a cold wind coming from the snow mountains.

The plane came down, in the late afternoon, on the Chiuchuan (Jiuquan) airfield outside the city wall on the edge of the desert. It was the end of our work in the air. It can truly be said that never before has such an elaborate aerial survey been made along the Northern Escarpment of the Tibetan Plateau at an altitude of less than 5,000 feet. One thousand miles of The Old Silk Road was always in sight. From now on for the next two months ground work would be the order of the day. We said goodbye to the pilots and mechanics who were returning next day to Shanghai with the films for development. In Chiuchuan (Jiuquan), a walled city set in the desert, there is dust everywhere. That evening saw the party exodus by 'hard' bus for a journey of nearly 100 kilometres to the oilfield at Lao chun miao (Lao jun miao). It was after ten o'clock before the bus arrived with a very weary party. It had been a fatiguing day in which much ground had been covered and a good deal of work done in aerial photography.

Next day, in conjunction with the VP of the Chinese Petroleum, the party laid plans for work in the initial stage. The geologists and geophysicists would work over the area from Yumen to Chiuchuan (Jiuquan). One other member would make a thorough investigation of production and refining methods at Lao chun miao (Lao jun miao). I would pursue an enquiry into the administrative side, in particular personnel, transportation, production (distribution of refinery products) and roadmaking.

The dreariness of the oilfield township was relieved today by a long trip on horseback along the valley of the Shih Yu River to the south. It is a gorge or canyon rather than a valley with gravel cliffs rising sheer on both sides to a height of 300 feet. The stream, running from side to side, is no more than ten yards at its widest. It is a torrent, in every sense of the word, and is icy cold. At no place is there even a semblance of a pool. The horses crossed and recrossed the stream and the water never rose above the stirrups, but the going was difficult due to the rocky nature of the river bed. We passed hundreds of small donkeys loaded with a special clay from the lower Nan Shan. This was required for making fire-bricks and utensils at the oil field and refinery. To control these trains of donkeys, making their way along the precipitous paths, two or three men followed behind each train. These men were queer fellows, primitive in every respect, almost black with the sun, their clothes were ragged and everything about them unkempt. Another interesting feature of the gorge trip was the number of springs welling up out of the gravel. It was obvious that the melting snows had percolated through the strata thousands of feet before appearing to gush out and join the river. Each of these springs was like an oasis in the desert, the surroundings were always patches of grass and green shrubs providing a relief to the rocky river bed and brown gravel cliff walls. At one of these springs, with its pleasant surroundings, were yaks and mountain sheep with their shepherds.

As I was involved only on the periphery of the party's geological and geophysical work little or no mention need be made of this here. Suffice it to say now that in the final report for the parent companies, to be completed in New York, I shall report under the following headings:

i) Inventory of present facilities at Lao chun miao (Loa jun miao) Field, including mechanical condition of equipment;
ii) Economics of Chiuchuan (Jiuquan) — Yumen Area outlining labour conditions, statistics, wages paid, subsidies and general conditions;
iii) Refinery — distribution of products;
iv) Weather, health and social amenities;
v) Military and political notes.

We are now in Central Asia, a part of the world where few have the opportunity to visit with the exception of travellers, explorers and a few missionaries who have a special calling. All the party are keen to see not only places and people but regions they have only read about. My diary, written at the time, plays an important part from here on so I will use headings as an introduction.

Kazakhs: July 24 1947

Journeying between the western edge of the Nan Shan mountains sometimes known as the Chi Lien Shan (Qilian Shan) and Lao chun miao (Lao jun miao) the party came across a nomadic band of Kazakhs. They were moving west but at the time when the encampment was sighted their tents were pitched on a grassy ridge where, strange to say, numerous springs bubbled up affording excellent pasture for the herds of animals belonging to the nomads. There were six tents, each about ten feet in diameter, with walls 4'-6' high then the canvas tapered to a cone by a most ingenious arrangement of rough woodwork easily folded inside. The Chinese were quite afraid and advised the party to pass by at a distance. For some reason or other the Chinese regard the Kazakhs as bandits.

In our approach, the nomads proved very friendly and invited us to look into their tents, in which the Chinese had previously warned us that many weapons would be concealed. Nothing however was visible and it was difficult to see where any dangerous weapons could be hidden. When it came to our cameras they were very curious and eagerly looked at the sights calling children and others around to share their curiosity. A cigarette obviously did not mean anything to them. Even when lit they did not know what it was all about. Their features and height suggest Caucasian type. They are short, on average 5'6", and broad-shouldered with a westerner's eyes and dress in many ways similar to the Russian Moujik.

The most outstanding thing about this small band — there could have been no more than forty including women and children — was the extent of the herd they owned. There were sheep, goats, horses and camels numbering about six hundred. Their speech is a dialect of Turki and they are notorious as raiders. Though very strong they are also lazy and apart from tending their herds usually have no other occupation. It was amusing and interesting to watch the women making ropes from animal hair. The dexterous use of fingers and toes in carrying out this work without the slightest hint that a strange eye was watching was in itself fascinating.

Lao chun miao (Lao jun miao) — Yumen Section

We are now on our way west to see the Caves of The Thousand Buddhas (Chien Fu Tung) regarded as having the most extensive collection of Buddhist art in the world. Coming down quickly from an altitude of 7,000 feet to 4,000 feet in a short space of time the change in temperature was noticeable and with the sun shining from a cloudless sky those on top of the truck soon realized sunburn was inevitable. But for the big oases of Yumen and Ansi (Anxi) the surroundings were entirely desert. So it was easy to maintain a surface on the Old Silk Road on which we were now travelling. After fifty miles we halted at Yumen for a noon meal. The 'fan tien' (place to eat) was a tumbledown mud house with a straw roof. The fare was rough and the place difficult to imagine with its untidiness and filth. However, the 'laopan' (old manager) did not spare himself and what he provided tasted good and we were satisfied. On this section of the Silk Road or NW Highway to look even for a modest Chinese meal is to look almost in vain. Those who do not take kindly to living on the land should make other plans. These highway towns are few and far between and centre around mountain streams which are diverted in all directions to irrigate the section and consolidate an oasis.

The Black Gobi

The journey between Yumen and Ansi (Anxi) was characterized by black gravel on the great expanse of the Gobi north of the highway. The gravel on the Black Gobi is estimated to be of pre-Cambrian age. It was possible here, without much searching, to come across evidence that the sea had once been in these parts. Sea shells were found but unfortunately mine disappeared in Hong Kong later when they were on exhibition with other things. On the other side of the highway, looking south, deserted townships stretching for miles were a grim reminder that without life-giving water, or the means to get at it below the gravel, humans cannot exist.

The Ghost Hsien of Ansi

We were tired and thirsty beyond words when our truck at last passed through the main gate into Ansi in the late afternoon of 26 July. At one time this must have been a big city. As it is we were told by the magistrate in charge, whose authority covered city and hsien or district of Ansi, the whole, as far as the eye could see, was now a ghost area. Truly it was a pale shadow of its former self due entirely to lack of water. Thirty years ago, within the walls the population was 100,000, now, at the time of our visit, 3,000 only. One cannot help feeling there was a lack of energy on the part of the inhabitants. Certainly the government did nothing to keep the people on the land. Artesian wells are the answer. The water lies below a conglomerate of gravels over the head streams; from there it is driven underground.

Ansi to Chien Fu Tung (Caves of the Thousands) July 27

Camping in the schoolhouse at Ansi (Anxi) shed some light on the living standards of the community. We were glad to occupy the courtyard, sleeping outside; although the temperature was high, at 4,000 feet, there were no mosquitoes. Next morning at six am we were off en route to the caves ten miles from Tunhuang (Dunhuang). The early stages of this seventy mile journey left no doubt in our minds that the term 'ghost', when applied to Ansi (Anxi) Hsien, was apt. At least six ruined towns, with their crumbling mud walls lay stark and bare in the morning light. The road for the first twenty miles was very good. Thereafter, with the desert to the north and the gravel foothills of the Chi Lian Shan (Qilian Shan), sometimes known as the Nan Shan, on the south, windswept sand covered what there was of road surface, to a skidding depth which made our journey by truck a real hazard. (Refer to map of journey on page 128.)

The Oasis at Chien Fu Tung (Thousand Buddha Caves)

At long last, close to our objective, we were confronted by a strip of land which formed the threshold of the most wonderful and certainly the greatest of all rock temples in Asia — The Caves of the Thousand Buddhas. This ribbon of land had many poplars and willow trees as well as crops, vegetables and fruit. It was indeed a welcome sight, an oasis of popular imagination.

The Caves of the Thousand Buddhas also known as: Mo Kao Ch'ueh

We reached the caves on 28 July 1947 and were introduced to

our host Chang Su Hang who was in charge from the Fine Arts Department of Government. Chang's job was to record the art of the caves and to effect improvements around the site. He was a delightful person and spared no pains to ensure we were well looked after and also in our visit to the caves that all available information was at our disposal. The caves were cut out of a high gravel cliff in three tiers over a distance, along the cliff face, of fully a mile. It is a cliff of fine grain gravel conglomerate which lends itself to the evacuation of large caverns. With the desert atmosphere and the extreme dryness of the structure the artistic work inside the caves has retained almost a pristine freshness since 353 AD when the first cave temple was constructed. At one time, from the second to the ninth century, this place was not only regarded as a centre of Buddhism but also a place where men of commerce were wont to meet being on the main highway through Central Asia to Europe.

It is estimated that the library in the caves yielded no fewer than 20,000 books and manuscripts. These are, for the most part, in the museums of London and Paris and, to a lesser extent, in Peking (Beijing) and cover a wide range of subjects: Buddhist texts, Taoist works, Confucian classics, novels and records of poetry and even books on medicine. They were written in many languages, a number of which Chinese scholars were unable to translate, including Runic Turkish. The caves were at their zenith in Sui Tang and Sung Dynasties. The Tang period easily exceeding all others — 681/936 AD, in quality of statuary and paintings.

After his 'enlightenment' when the original Buddha, known as Sakyamuni, was spreading the message in NE India, during the sixth century before Christ, no artistic likeness of the Master or the many facets of his teaching existed. It was not until the time of Alexander the Great that a foundation was laid which established art in the Holy Land of Buddhism. The result of both sculpture and painting portraying the life of the Master was a tribute to a blend of Graeco-Buddhist art with the best in Northern India. It was this artistic combination which flourished around the birthplace of Buddha that the Chinese pilgrims carried with them on their arduous journey home over the roof of the world and the deserts of Central Asia to a resting place at Tun-huang (Dunhuang), the site of The Thousand Buddhas.

Although some of the past glory of the caves was brought to light early in the present century the real significance of these treasures was not apparent to the Chinese Government of the day until the nineteen twenties. Somewhat later Chang Su Hang, previously mentioned, recorded a noteworthy interpretation of the caves from the fine arts department of government. It should be mentioned here that Japanese scholars have just published five volumes in colour on Caves of China, much of which is devoted to the Mo Ko Ch'ueh at Tunhuang (Dunhuang). This is the name given to the caves by the Chinese and Japanese, more on this shortly. Roderick Whitfield of the Department of Oriental Antiquities at the British Museum has also just published *The Art of Central Asia* a massive four volumes, largely connected with a walled-up rock-chapel from which Auriel Stein obtained priceless treasures at the beginning of the century from Chien Fu Tung at Tunhuang (Dunhuang).

Here the Chinese influence was brought to bear on the Sakyamuni's teaching of the Theravada principle, replacing it with the Mahayana concept where disciples can come to the assistance of the wayward on their journey through life to Nirvana (Heaven). In a similar way Buddhist art from India, with its Greek influence, was modified to be more in keeping with that of the Middle Kingdom (China). No wonder this place turned out to be a milestone in the movement of Buddhism from India to China, Japan, Korea and Tibet.

The Final Episode on the Caves at Tun-Huang (Dunhuang)

I have an original rubbing taken from a famous monument at the Caves, during our visit and it is of sufficient importance to merit a place in my story. The heading or title of the rubbing, in Chinese Character, is Mo-Kao-Ch'ueh, the last romanized word means caves. The first cave excavated in 366 AD was given this name, although the monument from which the rubbing was taken is dated 1348 AD and is also inscribed with the names of the donors who obviously wished to associate the whole complex with the name used at their inception. It is important to note in this connection, that both Chinese and Japanese archaeologists used the title of this monument in all references to the Sacred Oasis at Tun-huang (Dunhuang). Not so their

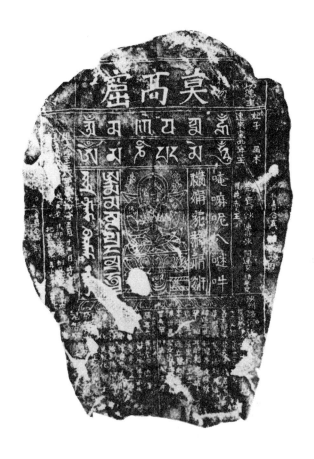

A rubbing from Thousand Buddha Caves

western counterparts for some unknown reason.

Be that as it may I now turn to the outstanding interest of this monument which carries an inscription in six different scripts. These appear in large characters above and to right and left of the Buddha image — see attached photograph of rubbing.

above	1	Sanscrit
above	2	Tibetan
right	3	Chinese

right	4	Hsi-hsia (Tangut)
left	5	Mongolian (Phagspa)
left	6	Uighur Turkish

Of the six scripts the chief interest is attached to the fourth — the rare Tangut script of the State of Hsi-hsia. This state was destroyed by Ghengis Khan about 1200 AD occupying as it did then the Kansu (Gansu) Corridor and the Tun-huang (Dunhuang) region in the eleventh and twelfth centuries. The inscription itself is the well-known Buddhist Mantra or Incantation:

'Om Mani Padme Hum'

To a Buddhist it contains a wealth of meaning. In point of fact a German scholar has written a book, translated into three different European languages, in which he treats the mantra in no fewer than four hundred pages.

Where mountain and desert meet

The great area of desert country, known as the Gobi, comes to an abrupt end in the region of Tun-huang (Dunhuang) where sand and mountain meet. The former stretching its long arm of sand into the foothills and the latter shedding its substance of gravels seems to result, at any rate hereabouts, in a victory for the Gobi. There is very little fascination about great tracts of sand but a walk over the desert here was surprisingly interesting. The large variety in shape and colour of the desert stones was a revelation. Three-cornered shaped pieces revealed the scouring action of the sand on two sides, the stone being pointed like an arrowhead in the direction of the prevailing wind. In colour it is possible to find several shades of red, black, green, grey, white and mottled stones in many hues. Quartz caught by the sunlight send out sparkling rays as if some precious stone lay hidden in the sand. All of this and the enormous dunes conspired to make this desert walk much more interesting than the prospect first appeared.

Standard of Living

A striking feature about all towns in Kansu (Gansu) is the poor quality of goods in the shops. In fact, the whole marketing set-up is of a much lower standard than, say, a Yangtse Valley town of equivalent size. The remoteness of the district is an important fact, but the purchasing power of the people and their consequent living standard gives an impression almost of poverty in all of western Kansu (Gansu). Ruined townships are evident and there are signs of a once prosperous countryside long since deserted by trade through lack of water. This prospect is in store for towns hanging on by the skin of their teeth unless a lackadaisical Government bestirs itself and attends to irrigation in its vast north-west.

Preparations start of the trip East by road: 1st August

Today was spent organizing equipment and preparing for the journey of some 700 miles to Lanchow (Lanzhou) via Shandan and Yongchang. There a lengthy halt will be made, at the latter place when the party will take to mule carts and trek into the Gobi to inspect an oil seepage area. Shortly after leaving Lao chun miao (Lao jun miao), following a month's inspection of the oil field there, and of possibilities in the area, a brief stop was made at Chiayukuan (Jiayuguan) where the Great Wall terminates, one mile above sea-level and 2,500 miles from the Yellow Sea (Bo Hai) in the foothills of the Tibetan Alps. Hereafter we came across many sections of the Wall in various stages of decomposition as we made our way along the 'panhandle' of Kansu (Gansu).

Fishy Passengers

Two trucks carried the survey party, stores, baggage, tents and satellites. The final categories deserve a paragraph to themselves. Our party included four guards armed with rifles and hand-grenades. But it is the satellites or 'fishy' passengers, as they were nick-named by the Chinese, that are our special interest here. They fell into three lots and preyed on truck drivers for one thousand miles along the NW Highway. i) 'Dry fish' were soldiers who forced their way onto the trucks and on finding a comfortable position relax and say nothing; ii) 'White fish', friends of officials who are put on the truck at stops along the highway where the vehicle must obtain a pass to proceed. This type of 'leech' pays nothing and actually issues orders to the driver; iii) 'Yellow fish'. They are as welcome to the driver as flowers in May. He collects from them to augment what he makes on 'pidgeon' cargo, already a considerable amount. My companion, in the last category, was a Jesuit Priest, Father de Brevery, formerly a banker in Paris then a Professor of Economics at Aurora University, Shanghai. What an interesting fellow he was! His story must wait for another time.

Chiuchan (Jiuquan) — The Wing Spring Town

Nearing the Wing Spring Town a flooded river provided a hazard. Already several trucks were marooned. Care in searching out the gravel bed and depth by sending out scouts to wade through, ahead of the trucks, enabled us to make the crossing. But it seemed advisable to remain in Chiuchuan (Jiuquan) awaiting flood abatement. What a blessing it was for me to have a day in this desert township. I recalled when we landed here from our chartered plane seeing a foreigner in the background and the somewhat patronizing way I approached him with a view to getting some political information. I was soon to learn that there is a code in Central Asia which enforces silence unless you are acquainted.

My Most Unforgettable Character

This is an account of the subject, which was written for *The Readers Digest*, and must be part of my story. It was not difficult for me to find the stranger with the reddish fair hair that I had encountered nearly two months ago at Chiuchuan (Jiuquan) airport. In the intervening period my attitude to the unknown had changed quite a bit but I welcomed an invitation from the stranger to see him on his home ground. His name was Eitzen — Father Eitzen — he came from Holland in the mid 1920s to work as a missionary in Turkestan later known as Sinkiang (Zinjiang). His original mission field was close to the Soviet border. In 1937 he was imprisoned by the Soviets who had taken over a sizeable area of China's Sinkiang (Zinjiang) province. During his incarceration of over five years with remarkable concentration and vision he had studied and mastered the varying dialects of Turki used by the Kazakh, Kirghiz and Uzbek as well as Arabic, the common language of the mosque in Central Asia. Russian and Chinese were just as essential to him as had once been the tongue of his native Holland and he was equally at home in English.

I was curious to see how the stranger lived. What work was he engaged in? And how did his standard of living compare with other oasis dwellers? Suspicious Chinese officials gave him an acre of Gobi wasteland on the outskirts of the town. It seldom rains there and water is the gift of the Tibetan Alps. In this place the right to water is more fundamental than the right to land and all must contribute their labour to ditch and canal water from the melting snows otherwise they had no entitlement to water. This was the situation that faced Father Eitzen five years before I met him.

Initially it was tough and the difficulties were almost insurmountable but having established a title to water and having assured himself that the allotted land was free from salty clay and unproductive sand, not uncommon in this area of Gobi, all was ready. The dry atmosphere and arid conditions rendered sun-baked mud bricks eminently suitable for building a protective wall and shelters within. The Reverend Father then set about making bricks and cultivating a vegetable patch.

An essential part of his day was set aside to attend to his own and the spiritual needs of the flock which he was, but slowly, gathering around him. Soon with loving care and the assistance of the faithful, when it could be spared from their own hard work, he built a chapel. In the eyes of Father Eitzen, there was no doubt, this was the House of God. He was so unassuming that I felt he had an understanding about the meaning of life far beyond that of his fellows engaged in the rough and tumble.

As we quietly walked from chapel to school and finally to a small clinic where he dispensed medicines received from his Order in the Provincial Capital nearly seven miles east, I looked expectantly around the small compound as I had seen nothing of a place where this humble man could lay his head. Finally I asked about this and he seemed surprised, explaining he had only recently finished the essential buildings. However, he was now thinking about a small place for himself. I doubt whether this ever got beyond the planning stage: we talked as we made our way through his sanctuary garden now rich with oasis fruit and vegetables and I could not help but feel I was in the presence of a man set by God to this place. And he was helping those around him to follow in the footsteps of the Master.

Sunlight on the Snow Peaks

This will always be regarded as one of the red letter days in our Kansu travelling experience.

The trucks left Chiuchuan (Jiuquan) at seven am on a perfect morning. At this early hour the view of the Nan Shan surpassed anything we had seen so far along the entire length of this great range of mountains. Wisps of white cloud floated like misty islands on the lower flanks, but the fascinating sight was well above 10,000 feet. The snow line and beyond, between fifteen and twenty thousand feet, glistened in the morning sunlight. These great giants on the edge of the Tibetan Plateau surely looked their best.

Ruined Townships

Nearing Kanchow (now Gaotai) the ruins of ancient townships were clearly evident. It is still possible to pick out original tiles used in the building of these 2,000 year old structures. It is interesting to conjecture on the reasons for so many ruined towns along the western section of this old highway. That they are so well preserved is undoubtedly due to the dry atmosphere. Regarding reasons for their being no longer part of the life in this province: in the first place the NW highway has long since ceased to be an important trade route; secondly, the lack of planned irrigation has driven mountain streams under the gravels. And finally, the attitude of indifference of the Central Government. Kansu (Gansu) which is only useful from a political point of view to be a wedge between the Mongol of Ninghsia and the Moslem and Tibetans of Chinghai.

Kanchow (now Gaotai)

In this province of occasional oases and arid desert the approaches to Kanchow (now Gaotai) were a revelation. From a pastoral angle the area surrounding this city in the pan-handle of Kansu is ahead of all others, the *raison d' être* being a range of hills to the north. These highlands trap and throw back the fast-flowing torrents from the Nan Shan mountains preventing valuable water from filtering into the Gobi and disappearing.

Shan Tan (Shandan)

The Great Wall is a feature of the countryside in this region. Although many sections of the mud defence have long ceased to exist, it is quite clear that in its erection thought was given to keep all pasture land and oases inside the wall. The object of our visit to Shan Tan (Shandan) was a stratigraphic one, to inspect the palaeozoic structures of the area in a further search for oil. Thin seams of coal of a poor calorific value were already being mined under frightful conditions. Shafts, unsupported, cut at an angle of 60° penetrating the earth about 100 feet made it possible to mine coal. Teenage boys crawl into these shafts where at the extremity a few boys only can work at the seams. Collapses are not uncommon, causing boys to be buried alive. Entire families, supported by these young miners, lived in caves nearby. Washing is seldom indulged in, making the whole scene along the coal face one that beggars description.

Yung Chang (Yongchang)

Our next halt was Yung Chang, on the highway, a drab and uninteresting place. Arrangements were made here for the hire of mule carts to carry the party with gear to Tsingtuching an anticline in the desert where there was an oil seepage. This structure was fifty miles from Yung Chang (Yongchang) and twenty miles from the last outpost on the edge of the Gobi.

Russians from Sinkiang (Xinjiang)

An interesting sidelight but one with political significance, crossed our path in Yung Chang (Yongchang). In the early hours of the morning the inn yard was invaded by new arrivals who spread themselves out without the comfort of shelter. In the morning to our astonishment we discovered the newcomers were Russians. All eight truck loads of them, 150 altogether, in family units. They were poorly clad, especially the women and children. The men on the other hand wore sheepskin coats and high Russian boots. The party was under military guard and had been almost a week en route from Sinkiang (Xinjiang).

Our Chinese geologists informed us they were Wei wu erh, the predominant native race in Sinkiang, now anti-Chinese. This was nonsense and surely they did not believe it themselves. Their speech was clearly Russian and local officials said they were going to Lanchow (Lanzhou) awaiting transport to Manchuria, their original home. This was also a thin tale. The real reason behind the exodus appeared to be that in all probability they were 'White' Russians who crossed into Sinkiang (Xinjiang) after the revolution in Russia and because of the present political position in that province are now regarded as a '5th' column by the Chinese Authorities.

The Gobi Village of Ning Yuan Fu

The first day's journey to Ning Yuan fu was most uncomfortable due to the rugged type of cart and the terrain. I walked about twenty miles of the way which was for the most part through desert country skirting mountain ranges to the south. After passing through the Great Wall our route lay along a basin where a clear stream, a rare thing in Kansu, provided enough water to make the valley a green and pleasant place to journey through. Ning Yuan fu the last village on the edge of the Gobi in these parts, was served by this stream which later disappeared into the Gobi itself. We spent the night at the village school, the surroundings of which enabled us to draw the conclusion that it was a prosperous village. By this time we were experienced judges of inns and camping places and considered this to be among the best of them. The supply of vegetables was adequate and cheap, as also were chickens and eggs. In fact, this village was the cheapest marketing place in Kansu (Gansu) as we saw it and our connections covered the entire length of the province: eggs were three for a penny and a chicken about ten pence.

With the object of reaching our journey's end some twenty miles further in the desert the caravan set out in the early morning over a rough desert track. Four miles from Ning Yuan fu we stopped at the last outpost, the house or castle of one Sun by name, in the hamlet of Hsipo. Ostensibly it was a farm, the owner being obviously rich. With only a few acres of arable land around his fortress and a supply of water difficult to hold against the thirsty capacity of the Gobi it was difficult to understand how he maintained these medieval surroundings. Part of the answer was that in the past a considerable trade was done with nomadic Mongols who came to the edge of the desert to carry on a barter trade with these outposts. This lucrative trade has ceased to exist, whether by compulsion or not, is unknown. All skins, camel hair and wool now move west from the Inner Mongolian Provinces to Outer Mongolia and the Soviet Union.

A description of this outpost is of more than passing interest. The inner sanctum, some seventy-five yards square, is designed in typically Chinese style, while the centre courtyard is protected by mud walls tapering in thickness from twenty feet at the base to twelve feet at the top, the walls being forty feet high. Four square towers are built in at the corners of the wall for the dual purpose of storing grain and providing look-out stations. Large rocks are also stored there for defensive purposes. A specially constructed shute guides the rock on its way to crush the enemy or undesirable who dares pass through the outer wall which is also a substantial barrier.

An Adventure in the Gobi

The next twenty miles of our journey was really a trek into the desert. We left the outpost at eleven am for what promised to be an uncomfortable six hours' journey if we elected to travel by mule cart. In actual fact it turned out to be ten hours. Standing out clearly in the desert, the shimmering mirage form of our objective, the structure where oil seepages had shown up seemed fairly close, but distances are deceptive in desert

Fort of Hsi Po on the edge of the Gobi Desert

country.

Rather than put up with an uncomfortable car journey over a rough desert track I decided to walk. Travelling light, crossing dry streams and sandy country on the way, I soon left the caravan behind. Rather weary and very thirsty I arrived on the edge of the uplift, the object of our trek into the Gobi. Now even on high ground it was impossible to trace any signs of the carts. I speculated on what would happen if the caravan did not turn up or if I failed to trace it. Fortunately, within the next hour two companions joined me having followed something of the same trail as I did. We soon traced the only water-hole, only to find it almost empty. The likelihood of what remained in the well of water being contaminated was uppermost in our minds. But we had an ample supply of Chloride tablets and decided to take a chance to slake our thirst. We saw no sign of the main party and with darkness coming down we decided to retrace our steps to Hsi Po, twenty miles distant. There seemed to be no choice after having observed, from a distance, a few strangers around the water-hole in the shape of wolves.

The night was pitch black and the going very rough. We stumbled into one dry stream bed after another and being very tired, hungry and thirsty the conditions we were in were not enviable. The outpost fort showed no light but we finally tracked it down to complete forty miles of walking in one day. The owner of the fort-cum-residence was extremely kind. After we managed to shake him up he soon had servants running hither and thither to supply our needs.

Finally we fell fast asleep, head to the wall, on a 'kong' (a bed with a fire below), in his guest room. In the morning we discovered, much to the barely concealed smiles of the servants, that when on the bed our feet ought to have been pointing towards the wall. Only when the human becomes a corpse is he placed on the 'kong' as we slept that night.

Return to the Fold: 8th August

Next morning we set out on camel-back to find the others, who must have been very anxious about our whereabouts. They had burned flares all night to attract our attention. When we looked at our terminal positions at the end of yesterday we were within three miles of each other. After the geological survey on 9 August the entire camp, about eighty persons in all including armed police and mule drivers, made an early start the following morning for civilization. The scene was reminiscent of covered wagon days.

Camels

For two of us it was our second experience of a long desert ride by camel, without a saddle. To say the least the travelling was rugged. Bactrian camels are quite different animals from the dromedaries of the Middle East. Their body is nearer the ground and they have two humps and as is common with all camels they know how to show their temper when loaded over what they know to be their capacity. In the Hsien of Yung Chang there are about 700 camels and their working period is between October and March. Freight rates are charged per day and it takes a caravan one month to cover 500 miles.

Bactrian camel the Gobi beast of burden

An Official Function at Yung Chang

The magistrate invited the party to breakfast. It was extraordinary to have such an affair so early in the morning. Never before have I sat down to a full-dress Chinese feast with the usual amount of wine and toasts at such an early hour. The leading question of the officials concerned was the prospect for oil. There is no doubt that a development of some kind in this area is urgently required to bring a measure of prosperity to their down-at-the-heels region.

The Pass of Maotaoshan

At the point where the pass is at peak the elevation is some 9,000 feet above sea-level. In many respects it is a dividing line between east and central Asia. The approach from the west might be regarded as the last of the cultivated country. The entire valley right up to the mountain top is green and if not cultivated it is used as grazing land. Descending to the dust country it is evident that the picturesque approach to the Maotaoshan is from the east.

Water Power

Hokou, or the river mouth, is at the confluence of three rivers. The rapid fall of the rivers to this point is perfect for making the flow of water into canals and as a result fast flowing mill streams. This is the perfect location for a series of water wheels set at intervals of half a mile over a distance of ten miles. The resultant power is harnessed to grain milling and irrigation.

Lanchow (Lanzhou) Water-shed

The country between Hokou and Lanchow (Lanzhou) provided us with a picture well worth recording. It was the last lap, but what a contrast between the Gobi and its oasis towns and that which we 'bumped into' now. Here the Yellow River and its tributaries predominated with a background of mountains covered with loess. These dust-covered mountains had been showing up for some time but now the covering of dust varied from a few feet to a hundred feet in thickness. On one side of the river, despite efforts of the dust to beat all comers, the never tiring Chinese farmer had scored success in carrying cultivation on terraces as high as 2,000 feet above the river. It was an extraordinary sight which made one wonder how water ever reached these fields.

Down at river level the Yellow River, now more than half a mile broad, was beginning to gather speed and volume. The familiar Lanchow water wheels, some of them forty feet in diameter, were situated at intervals along the river bank. Flowing at five knots plus the water impinges on the scoop-like vanes setting the wheels in motion. Approaching top centre the scoop pours the water into a trough from where it runs into irrigation canals. The final eight miles of the journey put Lanchow (Lanzhou) on the map as a fruit-growing centre. Between mountain and river on the left bank a two mile width of arable land was entirely given over to fruit-growing. Date trees predominated but apples, peaches, pears, cantaloupes and melons filled the entire area.

The Tomb of Ghengis Khan: August 17th

The remains of the Great Khan have been moved several times depending on political circumstances. When we visited the tomb it was at a place known as Shinglungshan, SE of Lanchow. The tomb is set in lovely surroundings. When the Japanese penetrated Inner Mongolia in the late 1930s the Chinese removed the silver casket containing the remains of Ghengis Khan and one of his wives in order to prevent the Nipponese from laying their hands on it and using the casket to rally the Mongol princes to their side. The casket has been moved yet again by the Communists, but where to is unknown. An oil painting of the terror himself, with piercing eyes and a flowing beard, hangs on the coffin.

The Kansu (Gansu) — Chinghai (Qinghai) Highway: August 18th

Indicators of a change in weather conditions were apparent. Overcast skies, followed by a dust storm which blotted out the city, boded no good for the morrow's long journey to Sining (Xining). When we set out early, heavy clouds hanging low, threatened expected rain at any time. The road was much inferior in condition to the NW Highway through Kansu (Gansu). Bridges were almost entirely absent. This meant that trucks had to go down into deep gullies whose steep gradients made descents extremely difficult and ascents impossible under conditions of heavy rain. Still journeying in the loess we left the Yellow River as its course turned SEE while ours followed the Hwang Shui River to Sining (Xining). The rain had now begun in earnest, turning the dust of the roadway into a complete slippery surface. It was easy enough for all of us to leave the truck and assist on a slippery incline. But it was quite another job to prevent the vehicle from skidding dangerously on a steep slope. Under these conditions and with the outlook all around threatening, we decided to seek shelter, awaiting a change, in a nearby village. Although the hospitality was cordial the shelter was poor. Some of us decided to sleep under the overhanging caves rather than risk the 'live stock' inside.

Gravel Fields

We were impressed en route by a considerable acreage of fields with a top surface dressing of large gravels ie stones of 3"/4" with a depth of an inch or more. In these gravel fields the moisture is retained longer in the ground and of course the gravel covering means that wind erosion is non-existent. Can you imagine the problem for the farmer ploughing a field with a gravel top surface?

Something New in the North West — Progressive Chinghai (Qinghai)

The crossing of a picturesque bridge high above a rocky gorge was our introduction to the Province of Chinghai (Qinghai). There is ample evidence of progress here. The roads are better and large gangs are working not only surfacing but building up rock walls to support the road. As Sining (Xining) is approached road surfaces are even better and we understand that Governor Ma is now engaged in constructing a highway across the Tibetan Plateau to the borders of his province and Tibet. Bridges and coverings over irrigation ditches, a constant source of trouble in Kansu (Gansu) show a marked improvement there. Perhaps the development that is most noticeable in Chinghai (Qinghai) is well planned afforestation. Such a scheme is calculated in the long run not only to prevent erosion but to make the Huang Shui River a clear stream. In the wide open spaces of the province agriculture takes second place to the rearing of cattle. The yak comes first in order of importance but the yellow cow, the horse, donkey, mule, sheep, goat and a small number of camels influence the economy to a considerable extent.

The Approach to Sining (Xining)

This huge province is only sparsely populated, 1,400,000 is the estimate, so that insufficient labour is a problem when it comes to work of a reconstructive nature. The Moslem element in the population is very great, in fact it is rare to find an eating place where pork is served; mutton is predominant. Nevertheless, we enjoyed the food and were agreeably surprised at the cleanliness of the kitchens and the presentation of the food. The valley of the Huang Shui was most attractive in the region of the capital. Willow and poplar surround the city for miles on all side.

The Founder of Modern Chinghai (Qinghai): August 20th

Our first duty today was a visit to the Tomb of the Governor's father. The elder Ma laid the foundation of modern Chinghai (Qinghai) with a policy of faithful lip service to Nanking (Nanjing). This involved nothing more than carrying out the orders of Central Government if it suited their own plans. Say what you like about the Ma family, they had proved themselves good administrators. The elder Ma, on whose tomb we laid a wreath, was responsible during his term of governorship for winning back to central control certain disputed areas of Tibet. The present Governor expressed himself anxious to interview the Party and in the meantime placed a secretary at our disposal. At three o'clock Governor Ma Pu Fong called on the party at the hotel and asked many questions on how to improve hie afforestation scheme and questions relating to problems on animal husbandry. Fortunately, two members had some experience in these matters and were able to offer advice. He was primarily interested in the work of our mission and particularly in the Hsiang Tan Minho area within his jurisdiction. General Ma finally asked for a personal report on the Min Ho oil seepage area. This request was backed up by the presentation of an unborn lamb fur coat to each member of the party. The skin on investigation by a furrier in Shanghai proved to be beyond treatment due to the original curing method.

T'a Erh Ssu Lamasery, or Kum Bum (in Tibetan) — The Temple of the Ten Thousand Images

Twenty miles south of Sining (Xining) stands the Tibetan Lamasery heading an imperative visit while in this region. It was a magnificent display of what remains of Buddhism as practised by the Lamas of Tibet. The golden-roofed temple buildings and the quaint dress of the Lamas is the first sight we had of this famous place which then was the residence of the young Panchen Lama. The trappings and symbols of Buddhism, in this, one of the most outstanding Lamaseries, was of great interest. Tibetan prayer-wheels and tapestries, together with a forest of bright curtains made the Hall of Classics something to be remembered. Two thousand Lamas could forgather here and kneel on cushions to be instructed on the high points of their faith in an atmosphere conducive to concentration. Apart from the temple itself there is a house of worship where penance and tribute alike are paid to the great Buddha whose image looks down on those who come to do him homage. The Buddha is surrounded by large brass lamps in the shape of a bowl, some a foot or more in diameter. These bowls are full of melted butter and the wick floating in the centre is always kept alight by a faithful Lama who, using a small vessel, pours butter fuel into the container and thereby feeding the flame. In the porch of this building facing the Buddha are pilgrims falling down on their faces and stretching out their arms in front of them. The repetition of this act of homage has over the ages made deep grooves in the shining wood of the porch. This strenuous form of worship can be maintained by the pilgrim for hours and often days. Other Tibetans could be seen arriving with tribute, in the form of grain, a much needed contribution, to feed the 3,200 Lamas of the monastery. One gets the impression that they would also benefit from soap and water. In fact it would appear, in themselves, they are sworn enemies of cleanliness. The temple itself creates an atmosphere of mystery, highly tinged with superstition. Grotesque animals look down on the courtyard, while the whole atmosphere is pungent with a heavy smell of incense. Prayer-wheels are spinning and a Lama with cymbal and drum is calling his fellows to worship.

The Panchen Lama

The temple buildings are situated on a hillside. There, on the upper slopes, a section is set aside as the residence of the Panchen Lama and his satellites. Passing through a moon-gateway, guarded by two armed provincial soldiers, we were escorted to the audience by a Moslem soldier, one of Ma Pu Fong's men who was to act as an interpreter. As we waited in the reception room it was observed that yellow was the outstanding colour though red was also much in evidence. Sitting facing each other in rows, with a centre passage between, the Panchen Lama when seated was able to look directly at us while we had to half turn our heads in his direction. He was attended during the session of introduction by three tutors, an evil looking trio, who constantly had their heads together and gave the impression that they had no little association with plot and counter-plot. The head of our party standing before the Lama and bowing presented him with a long strip of silk which was later returned as a memento of the occasion. Immediately after the silk ceremony each member of the party was presented to the Lama.

The boy Lama, twelve years old at the time, was discovered as the reincarnation of the former Panchen at the age of five and brought to Kum Bum. He looked bright and intelligent and will no doubt in time prove as capable and crafty as his predecessor. As temporal head of all Tibetans he has considerable power. We were later able to have a picture taken, the party and the Buddha in a group then each of us was permitted to take a picture of the Lama himself.

A Tibetan Village

The return journey to Sining (Xining) was made through a village in the vicinity of the Lamasery. In character it was essentially Tibetan, something quite new and, to us, novel. The quaint dresses and headgear of the villagers made them excellent studies for the camera while their close association with yaks and the high smell coming from both suggested they must surely occupy the same living quarters.

Sheepskin Rafts

I could not leave Lanchow (Lanzhou) without saying a few words about my reaction to a five mile trip on a sheepskin raft downstream on the Yellow River. The raft of inflated sheepskins, no fewer than thirteen, was held together by a light wooden frame. It was an experience not to be missed. I was surprised at the rapidity of this skin boat as it shot the rapids; not being rigid there was little or no sensation of shooting down. Each skin itself being a floating unit led to considerable flexibility in the water. Cargo rafts of cow skins can also be seen on the river, but my adventure was on a passenger craft.

All the foregoing concerning my journey to Kansu (Gansu) and Chinghai (Qinghai) is part of my diary, written at the time, and as such it is presented here.

Reporting on the Mission to NW China

On arrival in Shanghai from Lanchow (Lanzhou) in late August 1947, a few days were occupied advising the National Resources Commission of the Chinese Government and our own Principals of results arising out of our exploratory mission concerning oil possibilities in NW China. Thereafter the party left by plane for the West Coast of the States and then to New York where the results of our findings were discussed in much more detail. It is not my purpose to go into particulars on what transpired at any of these meetings. This also applies to my meetings in London and The Hague. Suffice it to say the subjects were to some extent political.

My reaction on visiting America for the first time since World War Two was surprise at day-to-day happenings and one in particular — the casual relationship in the office between the sexes. I will never forget my colleague, a very senior executive in Shell Oil, production, referring to his secretary, in her presence, as 'Sugar'. Don't forget I was four years out of the World War picture while in the hands of the Japanese. New York, then the hub of the oil industry in the USA took me out of my depth for a brief period, but it was all in the interests of growing up fast. It was part of this job for me to meet top management of all the companies represented in the Joint Survey Party. Apart from my own interest in the Exploration Trip I emerged with something new — an interest in geology and sadly, quite apart, the mess that China was in.

Together, as a family for the foreseeable future, after seven years of separation

Home again for a brief spell in the bosom of the family. I found them preparing, hopefully, for a return to China with me. The girls are growing up now, thirteen and fourteen years old. With parents always on the move their education has been in the hands of many teachers. Just imagine schools in Shanghai and Tientsin in China; Vaucluse, Sydney, Australia; Dumbarton Academy, Scotland; then back to Shanghai and finally Hong Kong. Astonishingly when asked to put their schools in a popularity league Scotland was away at the bottom, from June with emphasis. In the meantime Rena's mother had died and her father was living alone. It was a shock for him to see the family preparing for the long journey back to the Orient. Towards the close of 1947 we sailed on the *Empire Brent* from Glasgow, a most unusual port of embarkation, for Shanghai. Regular sailings to the Far East were not yet part of the shipping programme from the United Kingdom. Without a doubt it was the worst and most uncomfortable voyage I have ever experience and I am not referring to the weather. The accommodation and facilities on board were dreadful, to put it mildly, for such a long trip. But to be together was important.

Back in Harness

We had a lovely house in the French concession, Shanghai. I was in head office as a Divisional Engineer and for a period as Chief Engineer of the Shell Company of China. Shortly after returning in the early part of 1948 I made a business trip to Peking (Beijing) and it was obvious then that the communists, now firmly in possession of Manchuria, were preparing to move south. Before the war in the Pacific, Shell divided China, for the purpose of their business, into two companies. The territorial size of China made the Yangtse or (Chang Jiang), the long river, a suitable dividing line between Shell of North China, head office, Shanghai and Shell of South China, head office Hong Kong. Now, however, it was to be one company operated from Shanghai. Hence my responsibility covered all of China and Hong Kong. In the latter place a major installation was being built for storage and distribution at Kuntong near the Lyemun Pass entrance to Hong Kong harbour.

An on-site development presented the engineers with a problem. The chosen area, with land and water access, suitable for the plant was, in part, on a hillside. Much of this had to be levelled to accommodate oil storage tanks and buildings. At the outset a decision had to be taken on the amount of earth and rock to be removed to facilitate development. To cut a long story short following advice of surveyors in the Hong Kong government we worked on seventy-five per cent loose fill and twenty-five per cent rock. It was not long after starting excavation that the percentage of earth and rock proved to be in the reverse. Then the problem was how to dispose of the vast amount of rock occupying so much space that it interfered with development. A real Hong Kong problem was on our lap. It was finally solved by purchasing the biggest portable stone crusher from the US Army in the Phillipines. We were then in a position to see the last of the rock and sell it as aggregate to the fast-developing building industry in The Colony. The planning of this project and others of not quite such a major building task, all of which had to be approved by the Hong Kong Government, made it possible for me to meet their requirements as an Authorized Architect.

9

The End of an Era — Retreat from Shanghai

1948 was drawing to a close and it was obvious to all foreigners in Shanghai that the year ahead would see a major political change and with it the end of the International and French concessions in China's chief port. Let me turn to some of the happenings in Shanghai in the months prior to the 'revolution'. Aside from the actual political machinations, the outstanding indicator of downfall in this centre of trade was the day-to-day value of the Yuan or Dollar. This frequently resulted in a fall of five per cent in comparison with foreign currencies. Bank notes were printed daily showing astronomical figures. Business was to all intents and purposes at a stand-still. Stocks of oil were low but Shell was always able to arrange small shipments of gasoline and diesel fuel to keep the economy ticking over. The problem for management was what to do with stacks of Chinese dollars, the only form of payment. Shell handled this through purchases in bulk of rice, flour and such materials as had some value, even carpets. As their warehouse accommodation was empty, storage space was no problem. With little or no possibility of selling it, much was left and indeed must have been a prize for the Communists.

As a family, we were living from day to day, keeping our ear to the ground, so to speak. The girls went to school daily, thanks to Shell transport. We were living in a desirable area in the French settlement, so much so that we were warned and put under pressure from a very polite gentleman, who presented his card, and who advised us that we would have to pay him so much a week as protection money.

When the Communists captured Nanking (Nanjing) and were showing signs of making for the major prize Shanghai, Shell decided to get the Black family out to Hong Kong where I was posted as Operations Manager. What a relief, we made the journey comfortably by sea with only a few weeks to spare. By this time the Kuomingtang under General Chiang Kai Shek were on the move, via coastal ports, to Taiwan taking as much of China's treasures with them as their planes and ships could carry.

Hong Kong Mid-1949

It can be said without fear of contradiction that coming from Shanghai to Hong Kong at that time in political history was like coming out of darkness into light. Many Chinese with all their worldly possessions and who saw the writing on the wall were on the 800 mile journey south to a safe haven. The population of the colony then was just under one million, now it is five million plus, in thirty-two square miles of Hong Kong and 355 squares miles of New Territory.

The Blacks made the journey by sea and were delighted with their first home in Hong Kong facing the open South China Sea, about 1,000 feet above sea-level. My work was principally connected with the fast-developing new oil installation at Kuntong — an eminently suitable area on the mainland side of the harbour close to Lyemun Pass the exit or entry, by sea, to Hong Kong's land-locked natural harbour. Further responsibility extended to the coastal ports, the capital of Kwangtung (Guangdong) Province Canton and as far west as Kwelin (Ginlin) and Kunming in Yunnan. Time was running out for my job description to take care of mainland depots, but Hong Hong was bestirring itself as never before. I recall the visit of a Shell Director from London. While on our way across the harbour by launch, to see how the work was progressing at Kuntong, he said to me, 'What an enormous sum we are spending on this installation against a diminishing trade.' Little did he know, and at that moment he was perhaps unable to visualize, that there was a great future for Kuntong. Prior to the Second World War it was a belief held by many that Hong

Kong existed because, geographically, it was an ideal *entrepôt* for south China. Take away the need for its capacity as such and the Colony had no future. The Communists took the need away and Hong Kong prospered as never before.

Notable Events

HMS *Amethyst* a Yangtse River gunboat, the last of its kind in this waterway, was trapped in Nanking (Nanjing) by the rapid advance south of the Communists. It was very embarrassing for the navy as the captors had no intention of freeing the ship. The episode of how the Captain made his getaway of 100 miles to the estuary of the river and then to the open sea is well known. I recall the arrival of the *Amethyst* in Hong Kong and the great enthusiasm of all, with the White Ensign predominating.

The Korean War — North v. South

With the introduction, on both sides, of troops from far afield, Hong Kong was much involved in the movement of service men from the UK, Australia and New Zealand. Rena took part in canteen work which was essential for the men. It was an unusual sight to see so many uniforms, both army and navy, in the streets of Hong Kong.

Life in Hong Kong

In general, after Shanghai, life in Hong Kong was very pleasant and enjoyable. The climate is sub-tropical, much influenced by the land in the early and later months of the year, but the sea takes over, bringing lots of sunshine and humidity, from April to October with the odd typhoon playing a prominent part in June, July and August. Fine beaches are numerous so swimming is popular. But for us, with a fine launch at our disposal, water skiing proved the best of all leisure activities. Shopping was always attractive and Chinese service at home, particularly in cooking, added spice to life that made living worthwhile, not to mention favourite restaurants for our special choice. School for the girls was quite a journey, to and from Kowloon on the mainland, by car from our Peak home to ferry across the harbour, then by bus. This was the exercise getting to and from school which as far as I could see was enjoyable. We had a black spaniel dog, Duskie by name, which meant exercise for all over the many enjoyable walks around the well-kept roads and paths 1,000 feet above sea level. I had the opportunity of playing parts on China that were familiar to me, at the university. This was a pleasant pursuit and helpful in getting to know people outside of my business connections. As Operations Manager with Shell I was appointed Authorized Architect and gazetted as such by Government. As was the custom, following such an appointment, Rena and I were invited to lunch with the Governor and Lady Grantham at Government House.

On two occasions the family visited Japan. In these days of fast travel by air it was pleasant to go by sea. Our activities in Japan were confined to the main island of Honshu and seeing from on board ship something of the Inland Sea which separates the principal island from the small island of Shokoku. A shopping incident in Kobe will long be remembered. We were in a shop where antiques were the main attraction and Rena in a moment of enthusiasm picked up a pagoda-shaped vase for a closer examination. It was in three pieces and unfortunately the bottom section fell to the shop floor and disintegrated. From that moment the price, originally given us, shot up and we were in real trouble. Fortunately we were accompanied by a Chinese friend who, apart from being a Kobe resident, knew her way around. She is part of this story in the early days of war with Japan. Ai Lun Drakeford, Helen to us, said, 'Leave it to me.'

We left the shop and it was some time later before she joined us saying cheerfully, 'It's all fixed.' A number of Chinese girls have crossed our path during our domicile in East Asia and without doubt most of them were smart and very able.

A Visit to Macao

Portuguese Macao is only a few hours from Hong Kong by ship. On one occasion I had to make a trip there to purchase land for a service station. The only time I could see the owner of the land, to finalize a deal, was between two and three o'clock in the morning. He was an inveterate gambler and Macao was well known for its gambling dens. The landowner was quite willing to see me but his business hours were somewhat unorthodox.

Just before this time close friends of ours, the Nelsons, American missionaries mentioned in my story while in Hankow (Wuhan), were returning from Macao by air. While in flight the plane was taken over by gunmen and it crashed into the sea. Dan and Esther Nelson, with their teenage son and daughter, lost their lives.

A final word on Macao. During the mid-nineteenth century John Bowring, Governor of Hong Kong, on climbing the steep wide steps to the elevated St Paul's Cathedral, seeing the cross surmounting the façade he was inspired to write one of the great hymns of the age, 'In the Cross of Christ I Glory'.

The Union Church, Kennedy Road, Hong Kong

Kennedy Road runs parallel to the harbour and at an elevation of some 500 feet the Union Church is prominently situated against a rising wooded background. It was almost totally destroyed by the Japanese when they shelled Hong Kong from Kowloon in 1942. I was secretary of the Church in the late 1940s when restoration was almost complete. During my period in office I chaired a committee responsible for putting together a Constitution for Union Church.

Another matter, worthy of mention, came my way as Church Secretary. When the fire, resulting from the Japanese shelling, was extinguished the only part of the Church worth preserving was the massive oak roof members. This was quickly observed by the Japanese, after occupation, who were in need of materials to rebuild Government House which was badly damaged in their bombing of Hong Kong. After the war, British officialdom were delighted to take over Government House in better condition than when they left, thanks, in part, to the oak beams from Union Church. Our own restoration was in need of finance and as Secretary I was asked to approach Government on the subject explaining the position and suggesting compensation. I was not successful!

Travel by Air and Water

Towards the close of 1949 all four Blacks made a journey by air homewards using SAS. The plane was an 'Argonaut', famous for its breakdowns and we had some. When all were on board at Hong Kong we had to leave the plane again for twenty-four hours due to an engine problem and once more at Calcutta. Needless to say, our faith in this particular plane was at a low level. We left all the family well in Scotland and returned to Hong Kong by BOAC from Heathrow, not known by that name then, which was a mass of army huts. What a change today — thirty years later!

Our change of residence on arrival in Hong Kong was highest on the Peak, some 1,200 feet and with a beautiful panoramic view of the harbour — superb both night and day. The terminus of Peak Tram was close by so getting downtown was not a problem. Kuntong Installation was nearing completion and there was a lull in Hong Kong's post war expansion. We continued to enjoy living there for a further year before transfer to Thailand.

We sailed to Bangkok, complete with cocker spaniel, by coastal vessel. This way of travel is always good as it obliges enforced rest and as we moved South, sunshine as well. In August of 1950 Fiona left us to fly home to finish her education at Froebel College, Roehampton. While the parting was sad we were pleased that friends would meet her in London and she would stay with them, to begin with, a stone's throw from college.

10

Hong Kong, Thailand, and Countries of SE Asia

Living in Thailand after two years in Hong Kong was quite an experience. We were met on embarkation by a senior Thai from the Operations Staff. I learned later he was for some years at Manchester Grammar School, then Cambridge University, where he graduated with First Class Honours in Chemistry. Not content with this background, Boonyium finished his education at the Massachusetts Institute of Technology. His wife, by the way, was Minister of Education in the Thai Government. This was my introduction to Thailand. To this day he is a friend of mine. We stayed at The Oriental Hotel by the river side. It has changed much over the years but the site was perfect as an introduction to Bangkok. A feature of our hotel was a galvanized container on the veranda which served as a bath. I well recall the screaming of a daughter while in this so-called bath. One morning she looked up and saw a snake hanging from a tree close by.

Our first home in Bangkok was a very much down-at-heel Siam dwelling. Shell must have been going through a bad time when they leased such a place for their staff. The servants were Chinese and so different from our mainland experience. They had thieving down to a fine art. Even our clothing disappeared. Finally we decided enough of this and changed to Thai girls who were slow by comparison but housekeeping was a pleasant occupation thereafter. The Chinese in South East Asia is a subject that will be dealt with later.

Bangkok was tropical paradise. Canals of klongs ran parallel with many streets, palm trees predominated but the flame tree with its spread of beautiful red blossoms was best of all. A feature of all gardens and farms in the delta area was tropical fruit. To name one now, the pumelo is at its best when grown in a section of land where the tidal waters from the gulf meet the waters of inland rivers. The irrigation from this mix produces the pumelo par excellence.

The office was very much alive and prospects for our business looking good. The main storage plant and can-making factory was on the banks of the Chao Pya River, a tropical stream which divides Bangkok. The oil tanker capacity was limited to 10,000 tons because of river width and dredging of the sand bar at the river mouth, twenty miles distant. My job as Operations Manager concerned maintenance, budgeting and installation of future main plant needs. And throughout the country the economics of road, rail and water transport. Much travel was necessary, which pleased me, as I got to know the country and the people from north to south — 1,000 miles and east to west — 500 miles. These are approximate distances and of course, as the landmass narrows, in the long stretch of peninsula, the width is greatly reduced.

The People

By political edict, in the early 1940s, the name Thailand was introduced and replaced that of Siam. It was the land of the Thai people and the change made sense after over 1,000 years of occupation. The name Siam has its origins in antiquity and has no connection with the Thai people. The Thais were forced out of their homeland, Nanchao, the present day Province of Yunnan in SW China. Their exodus was slow but it accelerated during the T'ang Dynasty in China from 618 AD and Kublai Khan in 1253 AD exerted the final push. When the Chinese strength was north of the Yangtse River (Chang Jiang) there were periods when the Thais of Nanchao were very close to them. But now in their religion and language there is a strong Indian influence. In recent years the number of Chinese immigrants into Thailand, particularly Bangkok, has been considerable. When I am asked the question whether the presence or otherwise of Chinese in the community was advantageous or detrimental to it, I can truly say in my experience, from my economical point of view, it has been advantageous.

Travels

It is possible to travel expeditiously to Thailand using road or rail, by the former to even remote parts and by the latter there is a line from the border of Malaya to almost the extreme north of Chiang Mai. By air the link from Bangkok to north, south, east and west makes seeing Thailand for business or pleasure, comfortable. Given political stability I cannot imagine a more delightful place to retire than NW Thailand. The scenery and the weather combine to make this an ideal location to stay for a long time. On the other hand, the north east of Thailand is unattractive as a place to settle down. It is a poor country, uninteresting for the most part and a region where few of the country people are able to make ends meet, a real problem for central government! On the east it is bordered by Laos with the mighty river Mekong as the boundary. One day, hopefully soon, the river will be tamed to give hydro-electric power and all that goes with it to those, close by, who are so much in need. I was on the fringe of a study to achieve this end during my stay in Thailand. Regrettably, as is well known, this area is a minefield of political problems.

Away south, just 8° north of the equator, the island of Phuket, on the west side of the peninsula, was the chosen site for an oil depot so the Operations Manager set off to acquire the site and plan the layout. Phuket is a perfect tropical island, I can still see the tortoise making their way, if but slowly, up the beach from the sea in the early morning. Then above the high-water mark making a large hole in the sand to lay their eggs. The exciting thing about this manoeuvre, where mother laid her eggs in the sand, was the emergence a few days later of the young tortoise heading towards the sea. The pace was slow and they did not all make it. Birds of prey were on the look-out and many a young tortoise was picked up on dry land within a few feet of their objective.

An ideal place was found for a depot in a beautiful Phuket coconut grove. A point of some importance came to light when discussing the purchase of the land. Price was not everything, but the harvest of coconuts, for years to come, from palms still standing, after clearance of the site to ensure efficient operation, must be the property of the owner. Finally agreement was reached on this tricky point.

To the United Kingdom via Rangoon

In the Spring of 1952, with Rena and June, we set off home. The first stage was an hour's flight from Bangkok to Rangoon and then on the slow boat from there to Liverpool. I had a cousin in Rangoon, a resident of over thirty years, minus the period of Japanese occupation. He worked for a company who were about to cease trading. When the British Empire inevitably fell apart after the Second World War Burma decided to go it alone when they could very well have stayed in the Commonwealth. They were out on a limb and were in a sad state due to internal strife.

My native Dumbarton in Scotland had a close connection with Rangoon for many years. Ships were built there for the Irrawady Flotilla Company and largely manned by 'Sons of The Rock'. This name was given to local lads because of the Castle Rock which dominated the town. Burma Oil, a pioneer in the oil industry, who were the sole support of British Petroleum in their early days in Persia, had not a few 'sons from the same rock' on their staff. But when we arrived Burma was a closed book to foreign expatriates. It was a privilege for us to enjoy a few days in this delightful land to see something of the places we had heard so much about.

Selected on Rudyard Kipling by T.S. Eliot

> Ship me somewhere east of Suez, where the best is like the worst,
> Where there aren't no Ten Commandments an' a man
> can raise a thirst;
> For the temple-bells are callin', an' it's there that I would be

> By the old Moulmein Pagoda, looking lazy at the sea,
> on the road to Mandalay
> Where the old Flotilla lay,
> with our sick beneath the awnings when we
> went to Mandalay!
> Where the flyin' fishes play,
> An' the dawn comes up like thunder outer
> China 'crost the Bay!

On the way home from Rangoon by cargo boat we called at a number of ports. An unusual family incident at Port Sudan on the Red Sea is worthy of a place here. Before going ashore we were advised that all animals in these parts suffered from rabies. June and I planned to swim, the atmospheric temperature was at its Red Sea highest and Port Sudan is no place of beauty. Being an Arab country, the undressing places for male and female were quite separate. It was not long after we separated that I heard the most awful scream from June who had encountered a large tabby cat in her cubicle. The thought of rabies and being in Africa, the land of the tiger, was enough for June. It took us a long time to recover.

June worked with the Mercedes Car agency in Bangkok. This made it possible for me to get a price and delivery consideration for a car ex their Stuttgart plant. Both of us went to Germany for our first Mercedes. The service was excellent, their representative took good care of us. We saw the works and the Museum of Mercedes Cars and on seeing us on our way he asked of me, 'Is there anything I can do for you in the Federal Republic that you have not been able to handle yourself?' I quickly gave him the information of my search for the map made by Dr W. Filchner during his journey in Tibet (1926-28) never thinking I would ever see it. I was wrong — shortly afterwards he sent me a photocopy. June passed her Driving Licence Test in the Mercedes while we were in Crieff.

Our stay in Scotland was most enjoyable. We had a bungalow in Crieff for two months and Rena's father, who lived in Glasgow with his son, stayed with us. Crieff is ideally situated for trips by car to the highlands or lowlands. Picnics were a feature, shopping — for those who like it — excellent and the golf course was outstanding. What more could a family desire who were in need of a change from the unstable world of South-East Asia?

Fiona was with us too from her Froebel training college at Roehampton. As things turned out, Fiona put a spoke in the wheel of our joint return to Bangkok. She developed appendicitis and Rena had to stay in the UK for some time beyond my leave. June was now also at the stage when a college education was on the horizon. And in her connection it was Commercial in Glasgow staying with my mother in Dumbarton. June never liked either place but enjoyed living with Granny.

Bangkok Second Visit

The never-to-be forgotten BOAC 'Comet' carried me safely from London to Bangkok. It was a plane destined to make history in more ways that one. As the first passenger jet plane it made history. On the journey home, however, following my outward flight, a flaw in the body of the plane was disastrous. Nevertheless the 'Comet' was a trail-blazer in civil aviation and the Americans were quick to take advantage. Using the failure information from BOAC, Boeing and McDonnell Douglas modified their design and were first in the field for a long time, well supported by the enormous sum of money made available by the US Government for defence.

Back in Bangkok to a flat on one of those lovely roads with a 'klong' or canal as its centre and trees on either side of the water way. Shell were expanding their business and Thailand was a market for all their products which made the Operations Manager a key figure in the budget, planning and construction. This involved travel and I can say these last three years of my thirty years with Shell were the most interesting. I was fascinated by a number of everyday happenings which needs must be part of this story.

The movement of teak trees from the forests to the rivers using elephants never failed to interest me. Care of the animals and their skill in doing work which defeated mechanical transport was really remarkable in the observation of their antics. Floods from time to time were a hazard encountered in the north east but it was the ability of the people to cope and carry on that was a lesson to me. The cultivation of rice, particularly in the river delta areas, was something I had seen so often in China, yet here it was absorbing to watch the stage by stage growth of this plant which was the staple food of the vast majority living in our world. The procedure of planting in the muddy shallow water fields did not differ much from other rice growing areas in its back-breaking activity. It was the rise of the water in the paddy fields that held me spellbound. When and how long would it take for the frail rice plants to be swamped? It was intriguing to see the head of rice keep above the flood level until the depth of water was nearly one metre. In this veritable lake nothing could be seen but the heads, the grains of which were growing fast and appeared almost ready for harvest. As the water receded the rice ripened until the golden head fell flat on the mud, awaiting the farmer who cut off all the heads in triumph. As for its quality, when prepared for the table, it was the best rice I have ever tasted and from what has gone before in this story it is clear I have some experience.

Tropical fruits were in abundance, tasty for the palate and pleasing to look at whether in one's own garden or in orchards. The mango, pumelo, banana, mangosteen, custard apple and various types of melon were common, not forgetting the durian, so good to see, but with a very off-putting smell. Turning to flowers, Thailand is a country of the delicate and beautiful orchid. In my travels when I observed something new in fruit or flowers inevitably I returned with a plentiful supply.

After a month back in Bangkok Rena rejoined me, having made the journey by Danish freighter from Europe. Fiona had recovered from her illness and was back at college in Roehampton. On her arrival we took up residence in a lovely home with adequate grounds for a fine garden and pool. The latter was the result of the need to find earth for the foundation of the house. The water-table in the area was near the surface. The pool blended very well with the setting of house and garden, but it turned out to be the abode of snakes. It was not uncommon in these parts to find a snake coiled below a chair in one's sitting room. Quite often we observed an extraordinary happening from the porch of this house. During a heavy rain storm fish would suddenly leap out of the lawn surface to the delight of our servants who gathered them in readiness for their next meal. Again the reason could only be attributed to the high water-table. The coconut trees which were very much part of our garden attracted colourful birds and were perfect for the antics of playful monkeys. But in our memory the trees stand out, above all else, for a super dessert introduced by our Thai girl cook. Young coconuts were taken from the trees and she cut off the tops, in the same way as a boiled egg, poured off the milk and then into the interior of the coconut poured a mixture of beaten up eggs with a little sugar before replacing the top. The nuts were then steamed for an hour. Afterwards the taste of the custard impregnated with young coconut was out of this world, so much so that even our cocker spaniel Dusky came rushing to the table, saliva flowing from both sides of his mouth.

One cannot live in Thailand without coming in contact with the religion of the country: Buddhism. It is not the purpose of this autobiography to go into detail on the subject, even if I could. Suffice it to say that the South-East Asian countries are in the same school in their religious practice, though differing considerably from that of Buddhism in Tibet, China, Korea and Japan. Nirvana means heaven for the Buddhist and they cannot hope to get there until they have gained sufficient merit. This they cannot do in a lifetime here on earth. Therefore they expect to come back again to earth. This philosophy, as can well be imagined, has to be taken into account when talking to a Thai and why they are so easy going in their approach to life. So much of their time is spent on meditation and, when younger, many youths become monks for a short interval, before starting a career.

For me it was the art of Buddhism that proved so fascinating. My interest was focused on the Thai 'Wat' or temple and its surroundings including the wall which was always an architectural feature as it included an area of peace and beauty before reaching the 'Vihara' or temple itself. Thai ornaments are so alive, bold and flamboyant to be a special attraction in themselves. So much so they are regarded as one of the greatest expressions of eastern art. I refer particularly to images of the Buddha, for the most part created in bronze or stucco. It should be said that it was many years after Buddha died (543 BC) before art of any kind was associated with his doctrine. It was due to Greek influence, when Alexander the Great invaded India (356-323 BC) that Buddhist art first developed. There is an element of Green influence in Buddhist art, for all to see, in India and the countries of South East Asia where the Hinayana doctrine is practised.

A Foretaste of Nirvana by a Journey to Angkor and other Monuments of the Khmer Civilization in Cambodia

It has always been a longing of mine to visit that majestic conception of monuments in Kampuchea (Cambodia) planned and built by the Khmer under the direction of their god-kings and other high dignitaries. To me the key area of Khmer civilization lies south of the Dangrek Range and the basin of the Great Lake Tonle Sap. It is common knowledge that the Khmer monuments extended over a much wider area, even into other SE Asian states but my ambition was to see and appreciate Angkor and the surroundings, in the key area, of the pre and post Angkorian period (900 to 1200 AD).

Having obtained sufficient leave of absence from Bangkok we set off by car — Rena and two of our friends Jean and Edna. Our object was to do the journey comfortably in one day, from Bangkok to Siem Reap, some 250 miles. Alas we had a breakdown halfway and had to sit by the roadside for hours, until it was almost dark, waiting on another car — not an auspicious beginning. We arrived at Aranyaprathet on the borders of Thailand and Kampuchea as darkness took over. After dark one has to feel one's way around. Our only hope of accommodation was to bed down, all four of us, under the one mosquito net, on the floor of the local police station. Below us we could hear distinctly running water, it was a mosquito-ridden 'klong', just a slow moving stream of watery mud.

We were still 100 miles from the principal Khmer Temple of Angkor Wat. The town of Siem Reap, close by Angkor, was our objective. Here there was a hotel more in keeping with our life style than last night in the local police station. So much has been written about Angkor by archaeologists and others that I propose only to draw attention to highlights and monuments that impressed me. Many of the monuments were undergoing destruction through vegetation, but being a romanticist, not an archaeologist, I was thrilled to take photographs of the few monuments still overgrown with jungle.

Angkor Wat, like other Khmer temple mountains, was a replica in stone of Khmer cosmology. The central temple was Mount Meru, the pivot of the world, at whose summit lived the gods. The five towers symbolized Meru's five peaks. The wall enclosing the ensemble represented the mountains at the edge of the world and the surrounding moat, the ocean beyond. Who built these architectural symbols? A brief reference must be made here to answer the important question — who were the Khmer? Much has been written on this subject and one school of thought attributes their origin to India. For me, classic Cambodia has a primitive and aboriginal base. This is recognized when one eliminates, first all its connotations which are Chinese and second its Hindu cultural background. Khmer art evolved enriched by these two external influences and its own originality. It can there be said that the Khmer were an indigenous people reinvigorated by an admixture of Hindu blood and Brahman culture.

The bas-reliefs of Angkor Wat offer the greatest continuous expanse to be found in any existing monument. Those of the outer gallery measure over half a mile in extent, and those of the frontons, lintels, panels and other parts of the temple amount to nearly as much more. Aside from the shallowness of the sculpture, which often makes the feet appear in profile instead of in front and a few other similar defects, they are very close to natural. Most of the subjects are drawn from Indian epics and sacred books, *The Ramayana, The Mahabharata* and others, portraying legendary scenes from the lives of Rama, Krishna and avatars of Vishnu. The bas-reliefs held my attention more than any other of the many features in this land of temples. Finally, I would like to draw attention to a distinctive form of architecture in all Khmer Temples and Monuments, seldom if ever mentioned by archaeologists. For me, a former draughtsman, it was interesting to observe that Angkor Wat was perhaps the greatest of man's essays into rectangular architecture. In the overall research into Khmer civilization one finds an epic of rectangular forms has been imposed upon the Cambodian jungle. But it must be said the workmanship is classic in its simplicity and precision.

My work as Operations Manager with Shell, no matter where, was always paramount, very often outside of office hours, particularly when travelling. No matter in what part of the country one found oneself there was always an issue which dominated local gossip. It happened this way as my job took me to Ubon Ratchathani away in the east of Thailand, north of the Dangrek Range and close to one of Asia's greatest rivers the Mekong. The topic on the grapevine there, at the time, was

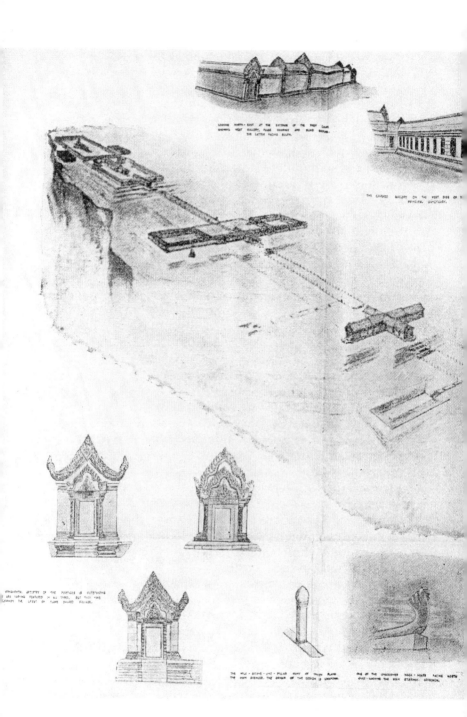

เขาพระวิหาร
Khao Phra Vihár

all about a certain temple fifty miles to the south of the borders of Thailand and Cambodia. Little did I think then that this monument was to play a prominent part in my leisure time as long as I was in Thailand and, believe it or not, for some time thereafter. The temple was 'Khao Phra Vihar' to the Thais and 'Preah Vihear' to the Cambodians. It was on record as a Khmer monument but very much neglected by both countries. Rumour had it that the position of the temple was now the subject of a boundary dispute. I simply had to see this place!

What follows, at some length, deals with how I became associated with this mountain temple. The book I wrote on the subject *The Lofty Sanctuary* of 'Khao Prah Vihar' published by the Siam Society and later quoted in the Court of International Justice at The Hague, must be part of this autobiography. Returning from Ubol to Bangkok by jeep I left the main road at Srisakes and headed due south to Kantaralak. Thereafter, the forest track, cut by cart-wheels, was rough and undefined. The rigours of the jungle trail, over the final fifteen miles, were softened by the beauty of the woodlands, the usual wild flowers and birds. At roughly 1,500 feet elevation the thick forest clears and you come out of jungle into daylight on a rocky plateau. From this vantage, out of the forest-clad slopes of the final rise to the summit, two large Naga (snake) heads in stone stand out. This is the entrance to the mountain ensemble of 'Khao Phra Vihar'.

Before proceeding further with a description of 'Phra Vihar' it is fitting here to quote part of the Introduction to my book by Prince Dhani Nivat, President of the Siam society.

'The approach to the subject by the author is that of the explorer more than the archaeologist. Mr Black has of course had wide travelling experience which entitled him to his Fellowship of the Royal Geographical Society of Great Britain. The attention he pays to what seems to be extraneous information in the form of legend and tradition deserves commendation. Legend and tradition have helped in the past to solve problems of scientific archaeology if examined in their proper perspective.'

The Lofty Sanctuary of Khao Phra Vihar

The passion for building sanctuaries on isolated hills is characteristic of the great religions of the east. Whether it be the ancient and mystic Hinduism or the gentle and peaceable Buddhism, one may see in many parts of the Asian mainland temples and sanctuaries on hill and mountain top. I should like now to take you to the most remarkable site for a temple in the whole of the Indo-China peninsula.

The length of the ensemble is 850 metres (2,805 feet) and it is composed of three courts with their entrance pavilions. The grandeur of the approach lies in a long steep stairway. The blocks of giant steps are recessed on their face. This simple ornamentation softens the hard effect of the stone mass and is known as the causeway of the Nagas, that mythical and semi-divine but graceful motive of the snake, used with great effect by the Khmer. The staircase as a whole achieves the grand effect of the heavenly approach. This Naga balustrade over thirty metres long (ninety-nine feet) is almost intact.

On the weather side, facing west, the first pavilion is largely in ruins. The significance of this building was possibly no more than a resting place on the way to the summit, but the decorative effect was not neglected. Much attention was given to the ornamentation of the doorway, a key point in Khmer sanctuary design. This pavilion is beautifully set as if on a stand in the form of a huge tiered foundation fully a metre (three feet) above the ground. This is characteristic of Khmer design and is a feature which reaches its crowning glory in the central monument of Angkor Wat which rests on a steep massive tier.

The long avenue between pavilions one and two has an easy slope paved with flagstones. Pillars lining both sides of the avenue, though now, for the most part, knocked down, must have added an imposing air to the approach. chiselled from solid rock and nearly two metres (over six feet) high, they were placed at three metre (ten feet) intervals along the avenue.

Before reaching the second pavilion a short pathway leads to a reservoir, hewn out of rock. Steps all round the rock have been made, not only to assist the water carrier but for artistic effect. The reservoir is a perfect rectangle known to the Khmer as a 'baray'. They were unsurpassed in the art of water conservancy and irrigation of rice fields. One thousand years ago three crops of rice, in a season, was common. Now it is a struggle to get one.

The second cruciform pavilion is a development of the first with additional sections on all four wings. The structure is almost intact but for the roof. The lintels and pediments in the building are a work of ornamental artistry. One lintel depicts a Garuda (divine bird with predatory beak and claws and a human body) on top of a Rahu head (monster who swallows the sun during an eclipse) above all is a four-armed figure of Vishnu (Protector; one of the gods of the Brahman divinity). The lintel and pediment is the major decorative point in all Khmer sanctuaries. The sculpture on these locations is as depicted in the sacred books of India. The first two pavilions or resting places are behind us as we traverse the avenue to Court III, a step nearer the summit and more elaborate in design. The entire ensemble here is best preserved of all and it is possible with little difficulty to construct, in the mind's eye, something of its former state and thereby get close to the glory of the past.

Along the much shorter avenue separating Courts II and III the causeway of pillars has an outer frame of a naga (snake) balustrade on either side. Much of the man-made setting has fallen, though by no means beyond restoration. The forest has taken over and there is almost an archway of vegetation between Court III and II. The entrance to the latter has partly collapsed, though there is still ample evidence of stone artistry on colonnettes, lintels and pediments, as described on our way through the lower parts of the monument. The two long halls or galleries constructed at right-angles to the entrance pavilion enclose the courtyard within which is the main hall and the east and west libraries. The first mention of libraries and books is in the early centuries of the Christian era, but it was not until practically the tenth century that the architectural innovation of the library appeared in Khmer design. It is certain that the two libraries in Court II at Phra Vihar were not built to take care of books. They might have been used for the safe keeping of astronomical instruments.

A continuation of the Main Hall in Court II leads into the sanctuary chamber and tower. This is the final stage and is in the form of a long hall with telescopic ends. At right angles to this hall on the east and west sides of Court I are covered galleries. On the south the Court is bordered by a blind pavilion which shuts off entirely the magnificent view over the lowlands of Cambodia. The principal edifice is set in the centre of Court I on the tiered mount fully a metre above yard level. At Phra Vihar the pyramid temple or Holy of Holies is intimately associated with the worship of the god-king, dedicated to Siva who is represented in the temple under the form of a linga or phallic emblem.

The Outlook from the Cliff Edge of Khao Phra Vihar, 600 metres above Sea Level

From the vantage point of the cliff edge looking south over the great plain of Cambodia there is a 500 metre drop which is almost breathtaking. To make a survey and carry out the photographic work for my book on Khao Phra Vihar — the high sacred monastery — I visited the mountain site on two occasions. I recall vividly the second time, following the rainy season. The days were beautifully clear, the visual distance was fully 100 kilometres (sixty miles). A carpet of forest green covered the undulating plain in the form of a vast area of vegetation at its best after the monsoon drenching. In the near distance villages could be seen in the middle or sheltering at the edge of vivid green rice fields. It was a view possibly unrivalled in its scope for sheer unobstructed vision. Although the tangled background of jungle romance is now largely stripped from the great monuments of the Cambodian plain it still lingers around Phra Vihar in its mountain eyre.

A lay-out plan of the temple and monuments of Khao Phra Vihar and two paintings of Courts I and II, pictured in their original state, are enclosed to help the reader follow my story.

Inscriptions on the Lintels at Phra Vihar

It is not my purpose to dwell at any length here on the subject of this paragraph. Suffice it to say in this considerable work it was my good fortune to have the guidance of that Master of Sanskrit and Khmer epigraphy, M. Georges Coedès, Past President of the Siam Society and member of *L'Ecole Française d'Extrême-Orient*.

No story is complete without a reference to the old man of the mountain, one Sri Kukhan Ketr, who lived in a cave not far from the stairway ascent to the monastery. He was a fund of lore and had an interpretation of his own for almost every lintel scene. Those who get to Phra Vihar should not fail to seek him out and take the opportunity of putting in writing some of his own and present day local interpretations about the mountain sanctuary.

Little did I know at the time when I recorded the work just described that my temple was to be the subject of a boundary dispute between Thailand and Cambodia at The Hague Court of International Justice, a few years later, more on this when the time comes.

As a family we were closely associated with the International Church in Bangkok. American missionaries organized the Church and their ministers conducted the services. A font, made in Hong Kong of Thai teak-wood, was presented to the Church by John and Catherine Black.

Our days in Thailand and indeed with Shell, were nearing retirement for me. Rena with YWCA connections was teaching English to young Thai students, and in this respect had to devote some time to the Thai language, which is a combination of Sanskrit and Chinese. It was a useful and worthwhile occupation. We will miss the long motor rides, at the weekends, into the country where we enjoyed watching the farmers in the rice fields and their activities during harvest. And added to this workaday job, the antics of those who climbed, with great skill, the coconut trees and showered the nuts to the ground from varying heights, but on average seventeen metres (fifty feet). Rena's father died at home in Glasgow during our final few months in Bangkok. We missed Fiona and June during our last term in Thailand but they were busy putting the final touches on their education prior to taking up school teaching and secretarial work.

Shell were very generous in giving me full retirement benefit at mid-fifty years of age. They took into account three years of incarceration by the Japanese. Through our Shell association we made a number of Thai friends and will miss them when we leave Bangkok but I am happy to say in the intervening period between the time of writing these lines and leaving Thailand we have kept in touch.

11

Leaving East Asia 'For Good'

The time had come to be on our way. Parties were a pleasant means of saying goodbye. We were ready for our journey on the good ship *Jeppesen Maersk*, essentially a cargo vessel with accommodation for only a few passengers. She set sail — a figure of speech — 27 March 1956 for ports in the Phillipines to pick-up cargo. The highlights from my diary, by date, will be used here, for the next forty-eight days, from Bangkok to San Francisco.

27.3.56

The big vessel finally headed down stream. The port is some twenty miles from the delta town of Packnam and dense tropical growth comes down to the river on both banks where houses on stilts are perched precariously on the muddy foreshore. Now the 'J.M.' (*Jeppesen Maersk*) was on her way to sea through the bar channel which is constantly being scooped out and sucked up by dredgers. One million cubic metres of silt is deposited in the channel every month.

28.3.56

The good ship 'J.M.' has forty-three souls on board; officers and crew with only our two selves as passengers. Her owner is Danish and port of registry Copenhagen but she never goes 'home', plying the world's oceans with 10,000 tons of freight, solid, liquid and refrigerated. From the eastern Atlantic coast of America, through Panama over the wide Pacific to East Asia is her 'beat'. From Baltimore to Bangkok is a far cry indeed but this is the 'J.M's' job. She carries the spices and primary products of the east to the west and on return manufactured goods. So much for the 'J.M.'

30.3.56

'Crumbs' dropped at the Captain's dining table today advise that it will be about six weeks before we arrive in San Francisco. We are loading for Vancouver too which promises to be interesting. This will mean the great circle route, skirting the Aleutians. The nights are beautifully clear and sailing conditions are perfect. The heavens are full of stars in a cloudless sky. Hereabouts, the Southern Cross and the Pole Star can be seen at the same time. The vessel is now roughly ten degrees north of the equator and the 'Cross' is well above the horizon.

6.4.56

Zamboanga, our next port of call, where the monkeys have no tails, is strategically situated on the main sea route from East Asia to Australia. It is on the narrow strait which separates the Sulu section of the Archipelago from the largest of the Phillipine Islands — Mindanao. The prosperity of the town rests on a coconut economy, indeed the town smells of coconut at every corner. A brief checkup on passenger ticket expenses while on board, all the way to San Francisco, worked out about three pounds per day. Where in the wide world can one travel with so much comfort under such delightful sailing conditions and be so well fed for such a small amount? The Maersk Line pick up eighty per cent of their cargo around the Phillipine Islands and it is by far the most valuable part of their freight.

11.4.56

It is a perfect day and this landlocked bay, where the ship is anchored, is like a sheet of glass. The loading of mahogany wood went on all night and thereafter we left for our next port, Cebu. Friends came on board then and we went off to a party. How pleasant it was to hear all about their way of life which was quite different from that which we experienced in Bangkok.

14.4.56

The ship arrived in Manila, a city of bright lights, just as the sun was setting. Our Shell friends came on board as soon as we berthed and whisked us off to stay in town. Manila is a very big city with confused traffic and many drivers in need of discipline. Nearly fifty per cent of the entire vehicular traffic goes by the name Jeepney. This is a converted Jeep with a decorated body made to carry ten passengers.

15.4.56

A run by car into the country — about forty miles — was very enjoyable. We travelled south and west to a high ridge at an elevation of 1,500 feet above sea level, to look down on beautiful Lake Taal with its volcanic island. The air was clear and cool and the view from this altitude wonderful. Back in Manila more hospitality was ours in the shape of a fine lunch at the Army and Navy Club, then farewell to the Phillipines. We boarded our ship shortly afterwards as she left for the China Sea crossing to Hong Kong.

17.4.56

A cold front from the land, about 100 miles from the coast, resulted in an unusual happening. Meeting the warm Pacific atmospheric conditions, a degree of humidity was created which actually made it possible to squeeze water out of our bedroom curtains. At this time of the year the south China Sea is the meeting place of air, which drifts over the cold land mass, and the moist warm winds of the ocean. The result of this makes Hong Kong notorious for fogs during April, May and June.

The wonderful approach to Hong Kong is of never-failing interest. The hills around the city of Victoria were covered in cloud but the early morning entry through Lyemun Pass has an appeal all of its own. This busy harbour and waterfront was an indication of Hong Kong's bustling prosperity. Despite this era of change, particularly in Asia, Hong Kong, even with its problem of surplus population, shows a sign of prosperity which is almost unbelievable.

18.4.56

We had lunch with a Chinese friend who deals a lot with mainland China. It would appear that goods are coming into Hong Kong — particularly food stuffs — unobtainable by the people in China. The Chinese on the mainland are suffering hardship due to lack of food. There is just sufficient rice but cooking oils are unobtainable. However, it is his contention that the lot of the common people, particularly farmers, has improved in comparison with Kuomintang days.

After a hectic day the ship sailed at 21˙.00 hours. What a wonderful sight Hong Kong is at night with its myriad of lights from the waterside to mountain peak. There was a thin veil of mist over the Peak itself but it did not take away the beauty of Hong Kong. As the lights faded out, on our way to sea, it was as if something disappeared with them — a last outpost of Empire.

19.4.56

There is now a full complement of passengers (twelve) so life on board takes on a different complexion. New faces and different points of view are always interesting. The east coast of Taiwan in preference to the straits between the mainland and the island is preferred because of the prevailing political situation. Our port of destination is Keelung in the north. The east coast is noted for its strong off-shore currents and being close to the continental shelf we are in the vicinity of the ocean's greatest depth. Normally the passage from Hong Kong to Keelung used the straits between the island and the mainland. The prevailing political situation ruled out the shorter passage, safety first making the decision.

20.4.56

We were condemned to stay on board and watch the cargo movement and shipping in the harbour of Keelung. The authorities refused us permission to go ashore. This did not apply to Americans who were welcomed with open arms but passengers of nations who recognized the Peking (Beijing) Government were not allowed to step off the ship. Taiwan, formerly Formosa in the days when occupied by Spain, Holland and later Japan, geographically is really part of China. The story of what is happening politically, about who rules the roost today, is gradually being unfolded.

23.4.56

During the storm last night a large flying-fish, eighteen inches long and eighteen inches over the fin tips made a miscalculation and landed on deck. According to the sailors, a light on deck was the attraction which took the fish off course. I must say, when the cook in the galley prepared this fish for lunch, it was in line with the best of its edible kind.

24.4.56

A very early morning arrival in Kobe's outer harbour brought with it the coldest atmospheric temperature of the trip so far. No need to remind ourselves it was early spring in these northern latitudes. When the good ship tied up at the pier it was comfortable and sunny but tropical wear had given way to flannel pants and a pullover.

Ashore we soon found ourselves in the famous Motomachi of Kobe. It is a lovely narrow street with all the dainty and artistic things made in Japan. On one side of the street an artist was carving bamboo panels with a skill which fascinated the passers-by. Here there was a shop with a display of lovely Nortake porcelain, there a cultured-pearl service station and close by an exhibit of exquisite lacquer ware. In a dinky little Motomachi tea shop, where tea cups have no handles and saucers are not used, there was a rockery in the corner taking up valuable customer space, but whose artistic setting, with rock plants and running water, was lovely in the extreme.

26.4.56

The highly industrialized Nagoya was uninteresting although we were advised it was the centre of Japan's porcelain industry. The good ship 'J.M.' left this tawdry port, where there is little or nothing to be seen from a visitor's point of view, at 7.00 hours the following day for Schmidzu.

27.4.56

The ship was anchored off shore and we experienced rain and more rain. It was a truly miserable day, so much so a number of passengers elected to go ashore and take the train to Tokyo.

28.4.56

It was a joy to see the sun streaming in through the cabin window just after daylight. The bay was beautiful and the green of the surrounding hills with some of Japan's best tea gardens and orange groves was a very attractive sight. The heights were still wrapped in white clouds and we looked in vain, before breakfast, for his majesty Mount Fujiama. It was not until after breakfast that Fuji-san, in all its grandeur, showed its conical snow-white head. Fuji is the highest peak along the volcanic chain of islands that form Japan. It rises to a height of over 12,300 feet and even in summer is snowcapped. Formerly an active volcano, it is now extinct.

Our ship loaded green tea, canned mandarin oranges, canned

tuna pike, oysters and plywood. It was strawberry time ashore and we did ample justice to this delicious fruit. By chance a restaurant was on our way which purveyed an excellent sukiyoki — a meal of egg, meat, onion, bean curd, young celery and a special sauce all cooked in front of you and eaten on a foundation of rice with chopsticks. The ship pulled up anchor and we sailed for Yokohama in the early evening.

29.4.56

This was our last port of call in Japan and the ship was 'tied-up' at a convenient wharf for going into town. The train journey from Yokohama to Tokyo, nonstop, takes just thirty minutes. We walked the length of Tokyo's principal street, the Ginza. As always shop assistants were so polite as to be embarrassing, but it was pleasant to receive courtesy in this day when it was fast disappearing. In Asia the essence of all languages is courtesy and respect. We invited friends on board in the evening and had the opportunity of finding out the whereabouts of mutual acquaintances and in general enjoying an evening that was all too short.

With the Azalea in full bloom and the floral display in shops arranged by the Japanese in their own inimitable style, all contrived to make our stay in Japan a very pleasant one. We slipped ropes from the bollards on the pier at Yokohama and with the aid of a tug we headed for the bay. The good ship 'J.M.' was making for the great circle route, across the wide Pacific, to Vancouver. The sun was shining from a clear sky. We could not have had a better day to set sail leaving Asia behind. The snowcapped peak of Fuji-san, away in the distance, was our sayonara to Japan.

1.5.56

It was May Day and we were heading north and east to the date line. Already the ship is some 500 miles out in the Pacific with dark brown sea birds following in our wake. It is said that these birds only keep us company as far as the 180th meridian, after that another lot take over. Sounds like an old sailor's tale.

On the great circle route the helmsman is watching his compass with orders to steer ENE. The sun rises at 4.30 hours and sets about 19.00 hours. The air is cool to cold and a fresh breeze is blowing on deck. After two years in tropical Bangkok, this is really a change.

3.5.56

A fresh wind today, from the east, has whipped up a moderate sea. Nevertheless, the ship is as steady as a rock aided no doubt by the fact that she is drawing thirty-one feet of water — the vessel's draft. Such a bulk below the water line fits in with the old sailor's maxim, 'steady-as-she-goes'. However, be that as it may, if the weather gets really rough this will be a very wet ship. The finest form of enforced rest can be had on board ship and if recuperation is needed there is no better place to find it. Our Captain is very hospitable and, as is common with seafarers of his ilk, who carry the responsibility of his command, he is an autocrat and is quick to lay down the law on almost any subject.

4.5.56

The weather has moderated and there is little or no ship movement. The direction of the ship is still north of east and from a landsman assessment we are south of the great circle route. If this is so we will have a greater distance to cover and be longer on the way. The Chief Engineer has invited me to see something of his diesel and auxiliary machinery. I shall take a few notes and make some calculations of horsepower from combustion chamber indicator cards. It will be a good opportunity to make a brief report on the Chief's reaction to the use of Shell Diesel Oil and Caltex Lubricating Oils.

5.5.56

This is the day the ship 'J.M.' crossed the 180° meridian, just after noon, according to the nautical 'boys'. This mythical line, like the equator, is the time for a joke — do we go under or over it? One fact stands out, this day Saturday, 5 May will be repeated again tomorrow and for all who cross the date-line.

5.5.56 (Second day)

Last night's party was a big success. The Captain was in top form and was the life and soul of the party. A ships officers' band contributed to the dancing success. An additional feature which gave the dancing an unusual angle was the sudden storm sea conditions which caused the good ship 'J.M.' to roll, so much so that the step of some dancers was a little unorthodox.

9.5.56

The intervening days were stormy and a change of wind to the north made the thermometer fall and there was, for the first time, a need for blankets on the bed. We are now within thirty-six hours of the Canadian coast, sufficiently close for our cautious Captain to announce that subject to weather we should reach the pilot at the Straits of Juan de Fuea early Friday morning, today being Wednesday.

10.5.56

The weather continues to suit the passengers but most are lying-low after the Captain's very sumptuous and hearty farewell dinner. This is customary at sea and Scandinavian skippers are well-known for their hospitality on such occasions.

    The sight of the first shore lights, just after dark, brought everyone on deck. Following eleven days of endless sea and having sighted nothing but birds and jelly fish, the coastline was indeed welcome evidence of civilization. The good ship 'J.M.' arrived in Puget Sound in the very early hours of Friday morning.

11.5.56

We passed Victoria, unfortunately, in darkness. However, by Namimo, it was early morning light and in its island setting, the town looked very snug rising steeply from the waters of the sound. Chilly conditions prevailed. The nearby mountains of the Selkirk Range were covered in snow while the Rockies beyond were not visible.

Vancouver is a beautiful city, a very desirable place to live in and although our stay was short it left a deep impression. Apart from its setting it has two fundamental assets which stay with me — the facilities of water and power. The latter through natural gas piped over the Rockies over Alberta and the former having the capacity to supply ten times the present populations potable water with no need for filtering or chlorination.

12.5.56

We are back on the ocean again with a long swell coming from the west. The sun was shining and it was not uncomfortable to be on deck as Canada faded in the distance. At Cape Flattery lighthouse, our first point of contact with the United States, the ship altered course 90° and we were soon on the way south to San Francisco.

13.5.56

If we had been asked to order the weather for this our last day on board we could not have chosen such perfect conditions for sailing. Rena's summing up of the voyage and mine too was that it was most interesting and enjoyable. Now we look forward to dry land and the long journey across America by Greyhound Bus. Soon now the good ship *Jepperson Maersk* will be going through the Golden Gate bringing an end to our voyage of forty-eight days from Bangkok to San Francisco.

San Francisco to New York by Greyhound Bus

We left San Francisco on the morning of 15 May, 1956 heading due south for Los Angeles. It is not proposed to describe cities at any length on this mammoth journey by bus. Suffice it to say we stopped off every night and in places several nights so there was an opportunity to see what was going on and get something of the atmosphere by talking to people. In conversation with our fellow travellers on the bus, from time to time they were unable to understand us and on our part we experienced the same problem. After leaving Los Angeles we crossed the Mojave Desert and then over the Black Mountains, south of Las Vegas to our first and most exciting halt — the Grand Canyon of the Colorado in Arizona.

Our stay at Grand Canyon Village was a perfect introduction. Here it was possible for the eye to see the jagged land forms which revealed how rivers and erosion together with geological layers, stacked on top of one another, shaped the canyon with its awesome depth and width. Cascading rock walls plunged nearly a mile to the Colorado River. Need one say more! It was a never-to-be-forgotten place and fortunately we were able to journey by bus, with frequent stops, to give the travellers an opportunity to place on record with camera, for others less fortunate, to see the magnificent beauty of this masterpiece in natural engraving.

Now on to Albuquerque where we saw something of ranch life and cowboys dressed for action. Kansas City was our next overnight stop followed by Des Moines and then a few days in Chicago. Here friends showed us much of interest in the second city of USA on the shores of Lake Michigan. We enjoyed an excellent lunch too, at the University. Our next halt at Wooster, Ohio, involved leaving the Greyhound and taking to the railroad not because of any problem but there seemed a need to be in New York earlier than the bus as the Cunarder was unlikely to wait for us.

At Wooster we stayed a few days with a geologist colleague, one of our team from my journey along the Northern Escarpment of the Plateau of Tibet when we were in search of oil. Wooster is a small town where rural interest predominates. We were especially curious to see and talk to a colony of Amish, a Mennonite sect, originally from Switzerland, but now citizens of the USA. Protestant by persuasion, very efficient farmers, wanted no form of Church organization, rejected infant baptism and the taking of oaths.

Thereafter Rena and I were soon in New York by rail living in a comfortable hotel. The stay there was short as the Cunarder *Queen Mary* was ready to take us on board for the last stage of our long journey home. The voyage was without incident and lived up to the standard set by Cunard wherever they sail. At Southampton the boat-train to Waterloo Station, London, nearly left without us due to an argument with the Customs about a camera I purchased in Hong Kong.

Returning to 'arrival home' and days that followed before settling down to an occupation in a new job still to be found, we relaxed for a year or more to get used to life in the Home Counties near London.

Our two daughters were living at Friar's Style Avenue, Richmond, practically on 'The Hill' close to that wonderful view along a stretch of the Thames Valley. We joined them there for a short stay. Fiona was studying at Froebel College, Roehampton, for a teaching career while June was on the staff of Distillers in their Piccadilly office having completed studies at Glasgow Commercial College.

The accommodation for us at Friar's Style was makeshift but suitable until we found something permanent at Shepperton. Here we were in a mews flat for a year until August 1957. Mews flats were in the grounds of a 300 year old Queen Anne-style residence on which a preservation order was imposed. It was close to the River Thames and Shepperton then was an attractive village. Our neighbours were interesting and altogether for us it was an enjoyable settling-in year. Beside us we had a young family with father in military service. Fiona was godmother to their eldest boy. In later years 'Dad' reached the rank of Major General and was Knighted when elevated to Commanding Officer of Forces in the United Kingdom. Later he was in command of land forces in Hong Kong. Our landlord, on the other hand, the owner of the estate, who resided in the Queen Anne residence was a snob. Possibly because of his movie world background as a producer and close association with Shepperton Studios close by.

In August 1957 we moved lock stock and barrel to Esher, Surrey. In The Woodlands there we stayed a very long time in a bungalow with a super garden. Accommodation with three bedrooms and a well-designed lounge was just enough for two of us and the main line railway station to Waterloo was within half a mile. This latter travel convenience was to prove a great asset in the years ahead. Fiona was still with us then but June, our youngest daughter, had just married.

We had not yet opened our teakwood boxes packed in Bangkok. It might well be asked why teakwood, a most expensive hardwood grown only in Thailand and Burma. Briefly, it was a Shell concession to anyone who was retiring from a teak growing country. Be that as it may, we were at last getting down to unpacking bits and pieces belonging to another world but very much part of the Black family.

From that time on my interest in antiques centred around porcelain, paintings — in particular scrolls, bronze receptacles, furniture including inlaid lacquer and my *pièce de résistance* from the Tang Period (AD 618-906): A Dead Man's Mount. The model animal, forty-three centimetres in height, still in my possession, is in wonderfully good condition. It was sealed in a tomb, for 1,200 years. The horse, a model of the dead man's mount, was put in the tomb in case the owner should have use of it in the next world. Its colour varies from light biscuit to pale jaune de Naples while there are orangy-red splashes about the saddle.

My enthusiasm for antiques had its ups and downs. During the war from 1941 to 1944, when I had no control over our belongings, which were in storage in a Shanghai warehouse, the Japanese made a number of my prized possessions disappear and, truth to tell, in our travels, others also had an eye on what they saw. We lost in transit a bronze vase, with an exterior floral dressing, all cast as one piece and two ginger jars, thirty centimetres high, in a brown and cream decor. Moving from one residence to another a dragon lamp-stand of hard wood, specially carved for us in Chungking (Chongqing) disappeared.

In my journey to NW China, after the war, while crossing the Black Gobi sea shells were found. I was really proud of this discovery and looked forward to the day when I could take the shells home and tell how they were found thousands of miles from the sea. After a lecture in Hong Kong University where the shells were part of an exhibit connected with NW China they disappeared. Alas, the opportunity of acquiring such a prize by stealth was too good to be neglected. I never saw the shells again.

Had the Chinese of antiquity developed their writing with a quill instead of a brush it is unlikely that the immense treasure of Chinese paintings would have evolved as it did. Painting and calligraphy developed hand in hand. For me the pleasure of viewing a Chinese painting includes enjoying the calligraphy of the written word, which is an art in itself.

In the last quarter of the twelfth century the Ma family in Hangchow (Hangzchou) were the greatest Chinese landscapists of all time. Ma Yuan (1190-1224) inspired the style of finger painting. The Mas were noted for their sharp drawing — every pine needle standing out. The Kano, an aristocratic Japanese school of painting, founded in the fifteenth century, was also famous for finger painting and indeed there are many museums in Japan today where this art is the outstanding feature.

I have in my records an excellent example of the Chinese finger painting entitled 'The Drunken Monk'. It is set in a woodland of pine trees and is associated with calligraphy in a composition of considerable power. Our home in Esher is furnished in an oriental style with chosen works of art and paintings as befits our background in the countries of East Asia.

12

Jobs and Travel following Retirement from Shell

Although I retired from Shell in my mid-fifties on full pension, a concession to those who had been in Japanese hands during the war, I was keen to find other employment. Shell offered me a job for a year, in their St Helens Court Office in the city of London preparing standards in text and drawing for their many overseas companies. This provided an excellent opportunity to keep a weather-eye open for employment associated with the oil industry. As a result of experience I was a Fellow of the Institute of Petroleum and it was through this background I found suitable employment.

About this time the whole complex of the marine world, as it applied to shipments of crude oil from source, was undergoing a major change. Briefly the size of tanker was increasing by leaps and bounds from 50,000 tons dead-weight to 500,000 tons. The term 'dead-weight' applies when the ship is designed to carry cargo in bulk. These giant vessels had to be loaded and discharged off-shore and while the economy of large shipments delighted the owners of the fleet it is correct to say that in the late 1950s there existed no port in the world capable of taking such vessels alongside a wharf. Obviously there was a need to give much thought to the problems involved, in the first place loading off-shore. When smaller ships were employed they anchored fore and aft and a flexible connection was brought up from the sea-bed, amidships. The problem with very big ships, however, was their surface of superstructure exposed to gale force winds and storm sea conditions. Anchors and mooring cables

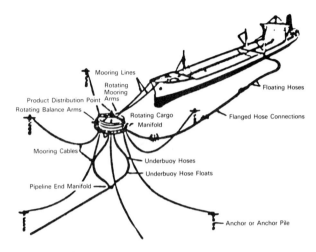

Catenary Anchor Leg Mooring

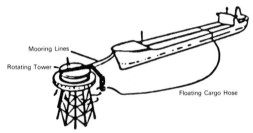

a) Fixed Tower Mooring

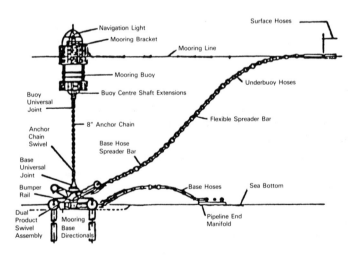

b) Single Anchor Leg Mooring

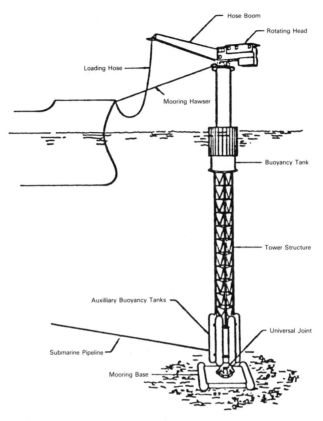

c) Articulated Mooring Tower

Sketches of Single Buoy Mooring for Tankers

could not take the strain so how was this major problem solved?

Undoubtedly the answer was a single point mooring at the bow of the tanker round which the vessel could revolve in any weather. The oil would be handled through a rotating manifold on a swivel arrangement built into the centre of the single buoy mooring. Having established the principle the nitty gritty was to make it work in practice. An experimental tank, about 100 metres in length and a vessel, to scale, moored to a single buoy were subjected to the maximum storm conditions. Meanwhile the ship's behaviour was closely observed. However, as I saw the complete picture a major problem existed, which the oil companies so far had given scant attention. The floating and under buoy, flexible hose connections through which the oil had to pass, while the ship revolved around the single buoy mooring, were of vital importance (see attached sketch).

After leaving Shell in London, in the autumn of 1957, I applied for a position with a rubber company through the Institute of Petroleum. American Rubber interests had recently taken over North British Rubber Company based in Edinburgh and were anxious to increase their business with the oil industry. To do this they were in need of someone who had oil company experience. The base of my work with US Rubber was their office in London but frequent visits to their factory in Edinburgh were necessary to get a knowledge of 'big' oil hose fabrication. And it was necessary, on my part, to draw the manufacturer's attention to off-shore site conditions under which the hose would require to operate effectively. The rubber companies had never contemplated such a volume of business and North British Rubber were ready to put their know-how to work so that they would be first in the field. Wisely, an early decision was taken to find an off-shore testing site. The Isle of May at the mouth of the Firth of Forth was ideal — get-at-able from the Fifeshire coast and with sufficient exposure to provide storm conditions for hose and ancillary equipment testing.

Unfortunately, at the outset, when super-tankers came into the picture, the oil companies did not fully appreciate the key position of the hose connection from sea-bed to below the buoy and from buoy to tanker-valve-manifold. I well recall my first experience when hose was involved and believe it or not the company was Shell. In 1961 Shell installed a single point mooring off-shore at Niigata in the Sea of Japan. As was their wont, the technician simply passed a requisition for hose to their purchasing department who in turn ordered from the rubber companies what they thought was needed as a result of their experience which in this case was nil. It should be said here the rubber companies were not inclined to challenge 'the might' of Shell's Purchasing Department. The result was and from a tanker terminal so far away, the first storm wrecked the hose lines and brought the operation to a stand-still. I visited Niigata shortly afterwards, that is another story but I can place on record here that the Marine Department of Shell, from then on, were very particular when it came to hose specification for single buoy moorings.

There was an enormous difference between working for Shell and US Rubber. Both jobs demanded and presented me with problems, some technical and others in the light of experience. Yet day to day discussions within the companies were as different as night from day. Briefly, when with Shell I was, in the first place, concerned with the operation of plant for reception, storage and distribution of oil products. Later the annual budgets for existing installations were involved and finally economic justification for the acquisition of new sites with the capital equipment required to make them operational. With US Rubber, on the other hand, sales talk was predominant except when visiting the factory then the operation of hose on Single Point Moorings at sea was discussed with a view to modification in design.

100,000 miles in 1963 to sell Oil Hose for Single Point Moorings

The following is quoted from *North British News*, on the above subject, in their May 1964 edition. When we learned of the vast mileage — equal to four times round the world which John Black of USR had chalked up in 1963 we asked him to record his impressions and what this aggressive selling means to North British Rubber.

> 'This sounds a tremendous mileage and I can assure you it is. Through fifteen countries about the same number of airlines, in planes from the old Dakota —

still a very safe machine — through turbo props to the latest jets 707s and DC8s with speeds of 250 mph to 800 mph. Perhaps the most frightening statistic is that during 1963, for 250 hours at least, I was on my "bottom" in a flying machine.

Your interest I am sure lies in what I was doing when not sitting in an aeroplane. Well over half the 100,000 miles was spent travelling to and from and actually in the area of East Asia. Going over the Pole in September of last year it was a great thrill to hear the pilot say, 'we are now abreast of the North Pole'. It was daylight and nothing could be seen below but a huge mass of ice with cracks in it.

The accompanying map gives you something of the coverage in the Far East. Indication is given of actual tanker terminals installed with Amazon (US Rubber Hose). I won't inflict on you the finer points of the economic study, sufficient to say there is an ever increasing tendency to build bigger and better tankers and this involves the same language in hose. At Port Dickson in the Straits of Malacca, north of Singapore, Amazon flexibles have discharged a 90,000 ton tanker with crude oil at 5,000 tons per hour. This ship is larger than the *Queen Mary*, nearly 1,000 feet long with 125 feet beam. Already plans are afoot to accommodate tankers of 120,000 tons at the same berth.

Going back to the map! US Rubber had orders for about £300,000 of oil hose for five major operations in this area during the latter part of 1962 and all of 1963. All of this made at Castlemills (Edinburgh), following a good deal of development work. As a specialist in this field I would like to say a good job was done, but it should be emphasized that the operation is a tough one and the best equipment is needed to stand up to the rigours encountered at these new off-shore tanker terminals.'

In addition to mileage shown on the attached map, the major oil producing area in the world — countries adjoining the Arabian Gulf — demanded attention, then, the Mediterranean area, particularly the North African coast. West Indies on the

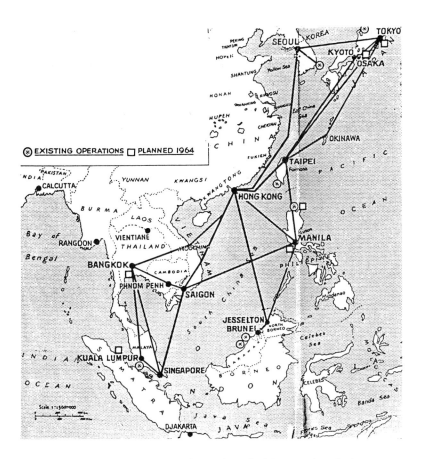

Map of travels in East Asia where SBMs were installed

other side of the Atlantic and as far south as Argentina were in need of being looked after at the request of users. To justify such a travel programme by way of results, there was a need to have a thorough knowledge, in the first place, of the delicate rubber hose connection from tanker to single buoy mooring. And, be familiar with the intricacies built into the buoy for the passage of oil and to enable the vessel to rotate round while at the same time transmitting the mooring load from a rotating arm on the buoy to the anchors or piles on the sea-bed. Such knowledge, on my part, was essential in advising possible users through lectures and being able to answer queries on this important development concerning the off-shore use of giant tankers.

No mention has yet been made of the USA and Canada in connection with Single Buoy Moorings for large tankers. Truth to tell, none of the ports on the east or west coasts were suitable for such giant vessels, even anchored well off-shore. My first acquaintance with the possibility of such a project in the USA took me to New Orleans and the Gulf of Mexico. I was in New Orleans a number of times lecturing on my favourite subject and to visit possible off-shore sites. The atmosphere in New Orleans made the visits enjoyable and I was fortunate to meet one of my fellow travellers who was in the team of geologists on our expedition to NW China in search of oil. Ned Clark, now Vice President of Shell Oil, very kindly put a helicopter at my disposal which made the job much more interesting. The Canadian experience came after I left US Rubber. Legal participation was involved through an accident at an off-shore oil terminal in the Bay of Fundy between Nova Scotia and New Brunswick. New York lawyers were handling the case; more on this later.

The tremendous change in our way of living after thirty years in East Asia deserves a word here. Don't forget my home for the first twenty years of life was in Scotland, now it is well south of the border. My mother was still going strong in her native heath, as was a brother and a sister, the former a schoolmaster in Galloway Country near Stranraer and the latter the wife of a Minister in the Church of Scotland. Doubtless it occurred to brother and sister from time to time that eldest brother was a 'renegade' Scot. As a matter of fact a job of work was the main reason for deciding to stay near the 'big city' and the choice of Esher in the lovely county of Surrey was perfect, with transportation advantage close to the main railway line for Waterloo in London.

Our youngest daughter, June, was married to a diplomat she met in Bangkok. Travel movements on business made it possible for me to see them, briefly, in places as far apart as Thailand and Turkey. But our most memorable time was the holiday Rena and I had with them in Israel. It was in 1960 when we made our first visit to the Holy Land. Tel Aviv was a good centre to plan movements and as we were staying with John and June whose work-a-day territory was Israel they were helpful not only in transportation but suggestions on what to see and the best way to go about it. After all our picture of Israel was a Biblical one and up to a point what could be a better guide than the Old and New Testaments. However, be that as it may, in the course of 2,000 years so much had taken place in the Holy Land we were in for a few surprises.

To start with our plan for sightseeing was easy going. Herod's ancient and splendid capital of Caesarea was first. Abandoned for over 500 years it is beautifully situated on the coast. Apart from the historical side and our interest in the aqueduct which brought water to Caesarea from the lower Carmel, believe it or not, we enjoyed a game of golf. Every blade of grass on the course was imported and in consequence the golf links was treated with 'loving care'.

In the centre of the country imagine a line drawn from Dan in the north to Beersheba in the south, roughly parallel with the coast. This places these townships, by the way, mentioned in the Old Testament, as historical border points in Palestine. Then to the east that great geological fault, the Rift Valley, originating in the Lebanon, demarcates the inland boundary of the Holy Land by way of the Sea of Galilee, River Jordan, the Dead Sea and the Wadi Arabah to Eilat. This encompasses the Negev Desert, now very much part of Israel. Away to the south on the coast we visited two of the five city states of the Philistines — Ashkelon and Ashdod. With the book of *Judges* in hand we recalled Samson was blinded hereabouts and later he pulled down the temple.

After seeing Beersheba we were invited into the Negev, by a journalist friend to call on a sheikh in typical desert surroundings. The encampment, where we were invited to have coffee, impressed us with the beauty of rugs spread on the desert floor. Indeed, the atmosphere and appointments in the tent suggested a good deal of comfort. In accordance with custom the opposite sex were conspicuous by their absence. The sheikh invited us to join him at their evening meal but we declined as graciously as possible. For us it was one more memorable day, of many that we spent in Israel.

'If I forget thee O Jerusalem: let my right hand forget her cunning. If I prefer not Jerusalem above my chief

joy let my tongue cleave to the roof of my mouth.'
These were the anguished words of a people in captivity. At last we reached the City of Peace, the centre of Christendom and without doubt the greatest medieval city in the world yet which has been destroyed many times. At this milestone in my narrative I resist a desire to write an essay on Jerusalem. However, I must set down some of the places where Jesus visited on his last journey to Jerusalem as these points are everlasting in my memory.

The Garden of Gethsemane

'And in the day time he was teaching in the temple: but at night going out, he abode in the mount called Olivet' (Luke 21). With *The Bible* in hand it is possible to follow the whole story of Holy Thursday night to the betrayal of Jesus. In the garden there are eight olive trees. Nobody can tell their exact age but botanists claim they may be 3,000 years old.

Dominus Flevit

A comparatively new church exists here built on foundations over a church of the fifth century. Its elevation on The Mount of Olives gives it a panoramic view through a number of windows of Jerusalem and Bethelehem. It was here Jesus predicted the destruction of Jerusalem and wept over it.

The Pavement

Last century the Roman pavement used by Pilate was uncovered. Here the Roman procurator set up his tribunal on a platform of pavement and pronounced his judgement on the trial of Jesus to the Sanhedron. The pavement is close to the Via Dolorosa. To this day markings can be seen indicating the variety of games played by the Roman soldiers. From *The Bible*, nineteenth chapter of John, therein is a vivid picture of the scene.

Hezekiah's Tunnel — 700 BC. Sometimes known as The Aqueduct

One day, at our most adventurous, we were bent on seeing something of a much earlier Jerusalem. In this connection we were well-rewarded by exploring the waterway of one Hezekiah (King of Judah, at the time of Isaiah the Prophet, see *II Chronicles* 32.30). This king recorded the making of Jerusalem's water supply system which was acknowledged as a tremendous feat in the well known Siloam inscription. The Assyrians were threatening the City of David whose source of water was the Gihon Spring outside the wall. Hezekiah camouflaged the spring or source of water and within the city itself excavated, through rock, a deep vertical shaft with steps to around water level. From this point a tunnel was excavated, again through rock, under the city to the water supply at Gihon. This great achievement allowed water to flow into the Pool of Siloam. The tunnel was our objective now, 2,660 years later. It was only knee deep and about 500 metres in length, but at the end we reached 'Cool Siloam's shady rill' my father's favourite hymn of a long time ago.

Some observations on Israel and their neighbours today

On a return visit to Israel in the 1970s my guide and friend while visiting Samaria was Professor Moshe Brawer of Tel Aviv University. When we reached Jericho our conversation centred on the problems facing the State of Israel with her neighbours. Out of all the angles discussed on the subject there appeared to be little or no solution though Moshe Brawer did say, 'I often wonder why the peoples of the nations concerned cannot get along together, when after all is said and done, we are all of the Semitic race?'

Perhaps those who are in conflict there today should pause and reflect on the question asked by Jesus, following the tolerant behaviour of the Good Samaritan, 'Which now of these three was the neighbour of him who fell among thieves?'

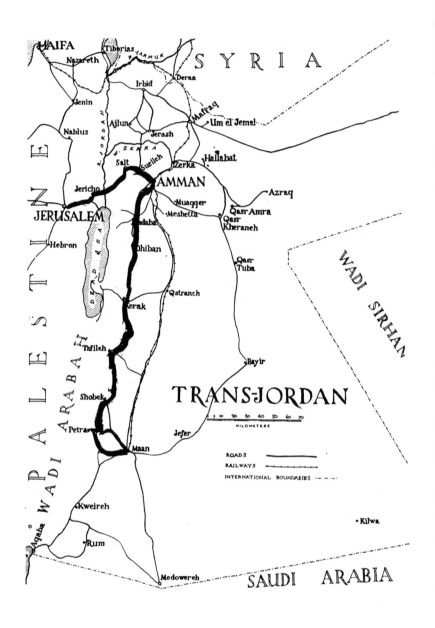

A Trans-Jordan Journey

Farewell Jerusalem! Now en route to the Rose Red City of Petra via Amman, East of the River Jordan

Seven of us, for the most part diplomats, set off in the early morning for our first break in the journey to Amman. A little time was taken off the route at the Dead Sea in getting to know something of the density while we were swimming. Crossing the River Jordan at Allenby Bridge was a landmark. The Capital of the Hashemite of Jordan was in a great state of excitement when we arrived. The King had been abroad for some time and had just returned to an unstable political situation. Soldiers were evident on roof tops with guns raised so our object was to get away from Amman with all possible speed. The route to Ma'an was due South, see attached map, and for the remainder of the day we needed all the hours of daylight to cover 150 odd miles to Petra. Alas, for this journey anyway, 'the best laid schemes o' mice and men fall aft aglee'. The party was frequently stopped along the way by armed guards who wanted to see passports and find out our destination. This was Lawrence of Arabia country, without doubt, on all sides.

It was approaching dark when we reached Ma'an, the nearest place on the highway to Petra. Here it was necessary to report at the local police station. This final stage of the journey was one of outstanding interest and the party were in two minds, whether to go on in the dark or stay the night at the police station. The question was answered by the ladies who were in the minority. They had just been to see what the police offered by way of accommodation and their opinion was — not for us, let's go on!

But what lay ahead? We hired horses at Al Ji village as recommended by the police. Each horse was accompanied by a guide, who was the owner, and as we left the village it was dark. One hour along a bridle path brought us to the narrow and romantic ravine of the Siq (the cleft). The classic name Petra and the early name Sela both mean 'rock', but one must go there to appreciate the meaning of the word rock, hereabouts. Entering this eerie gorge the winding path follows a dry torrent bed and above you, to a height varying from 200 to 300 metres, rise sheer cliffs. The night was moonless as we entered this strange somewhat awesome canyon. Before long there was a need for all seven of us to keep close together as the Arab guides had taken to 'pinching' the bottoms of the three ladies in our party.

After a mile or more of feeling every step of the way in this unreal world of towering sandstone cliffs we were suddenly brought back to earth with what looked like a large crack in the rock face. A move in the gorge at this point brought us face to face with what turned out to be an imposing tomb hewn out of the rose red rock of Petra, in late Greek style. It was midnight but fortunately for us the tomb face was lit up by acetylene lamps. Archaeologists were working on the facade of this great tomb — the Kaznet Far'on (The Treasury of Pharaoh). Of all the monuments in Petra this one is perfectly preserved. A little further on to the left there is a Roman Theatre cut out of living rock. We were now in the actual city, capital of the Nabataeans with its temples, palaces, baths and private houses occupied by them from the fifth century BC for 1,000 years. They were Arabic in speech, Aramaic in writing, Semitic in religion and Greco-Roman in art and architecture. When first heard of they were nomads. But by the third century BC they abandoned the pastoral for the sedentary way of life and were soon a highly organized society. Their metropolis Petra, which started as a mountain fortress, soon became a strongly fortified caravan station demanding safe conduct for caravans from as far away as East Asia en route to Europe via Gaza on the Mediterranean coast. One of their great problems was water and their extraordinary success in storing and transporting it was the most important factor in the prosperity of the Nabataean Kingdom.

At this stage in our exciting journey from Amman which started eighteen hours ago all of us were in need of rest and sleep. However, I cannot complete this tiring episode without reference to what happened at the 'last post'. Still on horseback we were riding along a Roman road terminating in a triumphal arch, beyond which was our destination. Alas we were not prepared for what faced us in front of the monumental gateway — 'Keep clear of the Archway, beware of falling masonry'. In view of this edict we had no other alternative but to get round the imposing arch. A steep descent to a ravine and a climb out were necessary. On the ascent Rena fell off her horse. Apart from fright, particularly in the dark of night, it was not serious, but she vowed — no more horse riding anywhere near Petra.

To complete this momentous sojourn in the Middle East, with

the Terra Sancta, Jerusalem and Petra now behind us, we simply had to see Byblos of Phoenician fame and the Bekaa Valley with its emphasis on the period of Roman occupation.

In 1960 travel between Israel and the Lebanon was impossible so Rena and I took a plane to Beirut from that part of Jerusalem which was then in Hussein's Jordan. We enjoyed the comfort and interest of staying in Beirut. The streets were filled by a motley crowd leaving a lasting impression of a city brimming with activity. There was much to see but our goal was Byblos further north on the coast. Byblos is a town as neighbouring Damascus, in Syria, is a city. They are the oldest continuously inhabited places in the world. Damascus comes into the picture later in this essay but now we are in the Lebanon.

It is true to say that civilization first developed in the Middle East, but that is a long story. Suffice it to say now that Byblos was the outlet for timber from the Lebanon to Egypt. And from the same source came papyrus (paper) from which developed manuscripts and books — hence the name Bible. It was however the inhabitants of the region, the Phoenicians, who left such a lasting impression. They traded throughout the Mediterranean and beyond the Pillars of Hercules. Alphabetical writing was the peak of their cultural invention. They were traders first and foremost and needed written document to conduct their business.

The ruins of Byblos are the focus of the place and thread of its history. The most striking and strangest of all monuments of antiquity, still standing in Byblos, is the Obelisk Temple, south of the Crusader castle. This temple contains the largest collection of stone gods in existence. Byblos to us was the Gateway to the Middle East and to history itself. The Crusaders were told to select ruined cities as sites as they would provide plenty of stones for castles which without a doubt they would need.

After Byblos we returned briefly to Beirut for a bus ride to Baalbeck. En route we entered a giant trough, running north and south, which traversed the entire country with mountain ranges on either side. To first time visitors, it was both impressive and strange. This was the Bekaa — part of the great rift valley which runs south by way of the Sea of Galilee, the River Jordan, the Dead Sea and the Wadi Araba below the Red Sea into Africa. In Roman times the Bekaa was fertile and a considerable source of grain, today it produces but a fraction of what was then required. For us we were greeted along the way to the valley of fruit and vegetable another way of satisfying human needs.

The Romans reached their eastern limit at Baalbeck. Here they raised a complex of buildings on a huge substructure about 300 metres in length. The columns of the Temple of Jupiter, on this site, are the tallest in the world and are said to be a quintessence of Imperial Rome. The Temple of Bacchus is well preserved and is probably the finest and assuredly the greatest Corinthian building of the entire Roman World. In the late summer of every year the Temple of Bacchus is the location for a musical festival. The complex of Baalbeck presents the amazing prosperity of the Roman Middle East and its high level of culture. It is sad to say, in conclusion, about this wonderful place which was a highlight of our visit to The Lebanon, that the famous cedars, close by Baalbeck, are now but a few in number. Even with reafforestation it will be a long time before the mountain recovers and regains its moisture-conserving mantle. It can be said without fear of contradiction that a traveller in the Middle East finds himself travelling in time as well as space. This was the experience of Rena and I in our sojourn just described.

I cannot leave this, what was for me none other than a pilgrimage in the Middle East, without a word about David Roberts R.A. (1796-1864). Roberts was born in Edinburgh and began his artistic career as a house decorator. Later he moved to London where Dickens asked him to design scenery for one or two of his productions, but he soon returned to scene painting.

In 1841 Roberts was elected a Royal Academician through his monumental original work on hand-coloured lithograph plates during his visit to The Holy Land, Syria, Idumea, Arabia, Egypt and Nubia. I have yet to follow Roberts' footsteps in Egypt and Nubia — more on this later. In the meantime, my record of the present journey includes twelve Roberts Prints (17" x 24"). A complete copy of *The Holy Land* in the original issue, Vols I and II; Vols III and IV Idumea and Petra; Vols V and VI Egypt and Nubia.

Roberts' own words speak for themselves: 'All I can say of the magnificent remains I have seen is that they are matchless.' And finally: Robert Fisk, *The Times* correspondent in the Middle East, not so long ago reported to his paper, 'It was regarded as prestigious in the well-to-do Arab homes to have a David Roberts' print displayed in a prominent position.'

Returning to our home at Esher there was time to reflect on our adventures in the Ancient Near East, the home of the first urban civilizations. It can hardly be said it was an exploration, although there was an element of this in our journeys. In this way we brought to light events of the past, making them more alive and above all finding in them eternal human values.

To work again

Back to the everyday world in London town. I was faced with different, but more agreeable, conditions of work. The International Division of American Rubber Interests with which I was associated decided to move their European Headquarters to Geneva. The reason for such a move was unknown to me, perhaps tax was involved. Their business in Europe was largely with the Oil Industry and London was obviously the place to be. At any rate for me it was a welcome change. I was appointed Manager of the International staff remaining in London and answerable to Geneva.

Someone in authority, to this day unknown to me, found us an office site in Brewer Street, just north of Piccadilly. It was a decrepit down-at-heel place where it would have been impossible to have business relations with our principal clients. A few words with Geneva and I was on the look out for a more prestigious office, which was found in the legal atmosphere of Chancery Lane.

In particular my work and that of my staff here revolved around the mono-mooring for supertankers. This single-point anchorage was then in its infancy with specific emphasis on the flexible connections from buoy to ship and sea-bed to under buoy. In most cases a unit capable of mooring ships of 500,000 dead-weight tons required 1,000 feet of large diameter hose. The oil was loaded and discharged at 500 tons per hour while the ship was safe at a berth in storm conditions.

Enough of this technical stuff and now to life in Esher and again in foreign parts.

In 1961 our eldest daughter Fiona married Sam Bradford. They left for Lagos, Nigeria, almost immediately afterwards.

Sam was a Chartered Accountant from The London School of Economics and in Lagos was in the advertising world. Experience obtained in darkest Africa was worthwhile, but it was not long before he decided to seek pastures new. On their first return to Europe a Canadian Chemical company offered an attractive accounting position in Fribourg, Switzerland. From then on to the present day, Switzerland, and our family living there, played a prominent part in our lives.

After our stay in Israel with June, our other daughter, and John they were transferred to the Embassy in Bangkok where I had the pleasure of visiting them while on a working visit to the Far East. Later, on return to the Foreign Office, John had an important position involving security at Heathrow Airport. He was ideally suited for this work which by all accounts he enjoyed. Thereafter, they were in Turkey and finally John was Commercial Counsellor to the High Commissioner in Malaysia at Kuala Lumpur. In my travels once again with a job of work in Singapore and Port Dickson I was able to see what life was like in the Malay Peninsula.

Our life in south-east England was pleasant and comfortable, neighbours were agreeable but we never really made any friends. Rena was on her own a lot yet managed to handle on my behalf all that crossed her path. When the Mole and Wye tributaries of the Thames overflowed in the autumn of 1967 boats were in use along The Woodlands and furniture with bedding had to be propped up off the floor. At this time I was out of the country and recall phoning from Milan for information needed from my files not knowing the state of affairs with Rena. Her reaction can well be imagined. Another problem left on her lap, I will never forget. I was then working from home as a Consultant. In this particular exercise with Imodco of Los Angeles, specialists in mono-moorings, I was responsible for a mooring off-shore at Chittagong, Bangladesh. This was a tricky job with tidal problems and required a Master Mariner to be permanently on site in charge. In this connection an advertisement in *The Times* seemed appropriate. I was out of the country shortly after placing the advertisement and Rena was inundated with phone calls, one as far away as South Africa. The mail was heavy and actual callers, on the door-step, even on a Sunday were experienced. I never ceased to be amazed at how she coped and what she did. The perfect secretary need not be your wife but mine was.

The Attraction of London

The appeal of living in Esher, only twenty minutes by train from Waterloo, was the come-on bait of what London had for me. (i) The Royal Geographical Society at Kensington Gore, where I was a fellow; (ii) The British Museum and (iii) The National Library, both under one roof in Bloomsbury; (iv) London University, close by the aforementioned, where The School of Oriental Studies was a treasure of information. And (v) last but not least, The India Office Library and Records, not far from Waterloo. Just what lies ahead in my biography records something of how I used these 'pillars of society'.

A pillar north of the Tweed, not mentioned above, The Royal Scottish Geographical Society, invited me in 1961 to lecture in Edinburgh on 'An Asian Journey'. Next morning, through the medium of a letter to the editor of *The Scotsman*, a lady sought the aid of some 'angry young Scotsmen' to take me to task for patting Buddhism on the back. The dear lady obviously did not know I could not talk about Thailand without bringing Buddhism into the picture. The incident brought to mind a problem of faith which is with me from time to time in the undermentioned prayer.

> 'Heavenly Father, help us to be willing to learn more about the world's religions so that we may understand our differences and share common convictions and go forward in faith to learn more of the truth as it is revealed in Jesus Christ.'

On visits to Edinburgh I had the opportunity of meeting the Bartholomew family of cartographic fame, particularly John who was then Editor of *The Times Atlas* ion the early 1960s. A letter received from him is attached which, to some extent, explains our connection.

JOHN BARTHOLOMEW & SON, LTD

The Geographical Institute,
12 Duncan Street,
Edinburgh, 9

November 1958

John Black, Esq.
Esher, Surrey.

    You were kind enough to give us your generous assistance in the preparation of certain plates of "The Times Atlas (Volume I): World, Australasia & East Asia", whether by the map material or the up-to-date information which you provided. We are pleased to send you, with our compliments and thanks, the printed copies of the particular plates with which you were concerned. We would like once again to express our very warm appreciation of your valuable help, which we know has contributed to the greater reliability and usefulness of the Atlas.

Yours faithfully,

John Bartholomew

Editor, The Times Atlas

Thoughts on the Past combined with the Present

It can well be imagined the way of life at home was for me vastly different from the forty years spent in a number of countries in East Asia. In adjusting to life in the English-speaking west I had three things very much in mind. In the first place to acclimatize my former way of living to that of a cosy London suburb. Secondly, as a mechanical engineer, in good health, I found an interest in off-shore engineering which called for travel. And finally the attractions of living near London, together with many notes in my possession, written during travels in Asia, enabled me to retain a live interest in the culture, religion and way of life of the peoples in whose lands I had lived so long.

Looking over talks I was invited to give on subjects of special interest, a lecture at The Centre of Far Eastern Studies (London University) was the most scholarly. In November of 1974 the Chairman of the Centre asked me to address his School of Oriental Studies on 'A Kansu-Chinghai Journey' (NW China) 1,500 miles along The Old Silk Road from Xian (Sian) to Tunhuang and The Caves of The Thousand Buddhas (Ch'ien-fu-Tung). Professor William Watson of The Percival David Foundation of Chinese Art was in the chair and I was faced with a formidable audience. The lecture is attached as an appendix to this autobiography together with a map of the journey and a sketch, at some length, with a description of The Thousand Buddha Caves.

The Royal India, Pakistan and Ceylon Society were next on my list of lectures in London. A journey to a Khmer Sanctuary in Cambodia, was the subject, with special emphasis on architecture from photographic records and drawings. This monument has already been referred to in my book on the subject published in Thailand and it plays an important part in my presence at The International Court of Justice, The Hague, where the temple was the subject of a Boundary Dispute between Cambodia and Thailand — more on this shortly.

French connection in Thailand and later in Paris

While in Thailand it was a privilege for me to know Georges Coedès, Director of the *Ecole Française* in Hanoi and former

President of The Siam Society. Coedès was a master of Sanskrit and Khmer epigraphy and the author of *The Indianized States of Southeast Asia*, sometimes referred to *India Beyond the Ganges*. The book was first published in 1944 and again in 1948, it was an acknowledged authority on the subject. When the author was resident in Paris in the early 1960s he was persuaded to undertake a revision of his master work. At Coedès's invitation I came into the picture, because of my background in China, to write a section on irrigation and transportation for the revision of his book *The Indianized States of Southeast Asia*.

During the early part of our era a people known as the Funanese emigrated from south-west China to the southern region of the Indo-China Peninsula. This river delta country, with its Lake Ton li Sap, an ideal reservoir, was perfect for them, with their China background, to develop agriculture. Where and how did the Khmer come into the picture? It should be emphasized here that the Khmer also had a Chinese connection. On emigrating they were within the cultural structure of Southeast Asia under the influence of India, a country steeped in Hindu and Buddhist tradition. These religions gave rise to the Khmer, achieving the mysterious and passionate beauty of the Stones of Angkor. Cambodia was named as the country taken over by the Khmer in the period 600 to 700 AD, but it was 500 years later before they reached the peak of their fame in temple architecture and the display of great mythical themes like the 'Churning of the Sea of Milk'.

I spent many hours in the British Library, Bloomsbury, and the India Office Library and Records near Waterloo in search of information on sites and records of irrigation methods used at the time when Southeast Asia was very much under the influence of India and China. On nearing completion in the mid 1960s Coedès introduced me to a colleague still active in the *Ecole Française d'Extrême Orient*, then in Cambodia. This man had recently been able to use a helicopter to take photographs at a low altitude in the Angkor area showing the irrigation layout with its Barays or reservoirs. He refused to give me any information or photocopies resulting from his helicopter flight. This put 'paid' to much of my work as it would have set a seal on what I had done.

The conclusion arrived at through my research was that the Funanese who originally occupied the territory, later conquered by the Khmer and known as Cambodia, planned and made practical a system of irrigation beyond anything the Khmer were capable of. To make a foundation for and to erect the sculptural marvels designed by the hierarchy of the Khmer required a population who were supported by a high standard in agriculture (three crops of rice annually). The Funanese brought such a standard with them from China. The great masters and George Coedès was one of them, who wrote innumerable articles on the Khmer Kingdom of the Cambujas, failed to recognize the important role played by the Funan who laid the economic foundation for the God-Kings of Cambodia.

13

International Court of Justice, The Hague. Case Concerning The Temple of Preah Vihear (Cambodia v. Thailand) June 1960

Reference has already been made to this temple during the time I was in Thailand under the heading of 'The Lofty Sanctuary of Khao Phra Vihar'. The detail therein will not be repeated. Suffice it to say a lecture was given by me on the subject to the Siam Society which was later published in book form and to this date is on sale from the Society. My object in referring again to the temple concerns a boundary dispute taken to the International Court of Justice at The Hague. When the Court was in session I was resident in the United Kingdom but they were good enough to send an invitation for me to be present at the hearings.

The dispute was brought by Cambodia against Thailand. The former claimed that the temple and site on which it was built was in Cambodia, but recently it had been occupied. To use the legal language for the 'Subject of Dispute': 'Notwithstanding the repeated protests, diplomatic representations and complaints of Cambodia the Kingdom of Thailand has, since 1949, persisted in the occupation of a portion of Cambodia territory, where there are ruins of a holy monastery, The Temple of Preah Vihear, a sacred place of pilgrimage and worship for the people of Cambodia, up to the present day. In order to ensure the respect of its rights and secure from the Kingdom of Thailand the fulfilment of its international obligations Cambodia has thus been impelled to bring the matter before the court.'

It is not my intention to repeat here what has already been said in my book *The Lofty Sanctuary of Khao Phra Vihar* with the exception of what comes under Annexe LXXXV in the official Hague court documents.

The Lofty Sanctuary of Khao Phra Vihar
par John Black
F.R.G.S

'Proceeding due east from the wing of the first gopura, a path is encountered. At first this is no more than a footpath, with fragments of well cut rock appearing above the vegetation on either side. Then, there is clear evidence that the path was once an avenue six metres wide and bordered by heavy sand-stone blocks. These were covered with moss but, when examined, were found to be blocks with a chiselled surface, laid one on the other. This avenue from the Cambodia lowlands was about one kilometre in extent and led to a steep spiral stairway solidly made in stone.

'On the northern flank of this now largely overgrown avenue the hillside falls almost abruptly into a large depression or basin before it rises again just as steeply to the rocky basalt plateau, where Nai Amphoe has so kindly built a rough sala for the traveller. But for the elephant tracks it would be difficult to wedge a way through the thick vegetation of this basin along which courses a stream. Local legend has it that this was a former reservoir. Making use of the natural depression, the builders are said to have converted it into a dam, by controlling the east outflow to provide the large water supply needed by thousands of workmen who must have been engaged on a task the magnitude of Phra Vihar. There was, however, no evidence that this natural depression had been used as such, though the job of creating a reservoir would have presented no difficulty to the Khmer.'

Reply of Mr Dean Acheson
(Counsel for The Government of Cambodia 22 March 1962)
Hon. Dean Acheson, Member of The Bar of the Supreme Court
of the United States of America

'To return to the question of stream channels, there is further evidence for the existence of the stream flowing through point F onto the Cambodian plain. John Black, a distinguished British archaeologist (an exaggerated distinction by The Hon. Dean), visited Preah Vihear in 1955. His article entitled 'The Lofty Sanctuary of Khao Phra Vihar', was published in Volume 44 of The Journal of the Siam Society, 1956. We have given the Registrar a reprint of this article.'

The article referred to comes under Annexe LXXXV already quoted.

'In other words Mr Black is saying that to the north of the east-west staircase he found a valley also running east-west. The existence of this valley is attested by all the maps in evidence and by all the testimony, written and oral before this court.

'Mr Black says further that following elephant tracks through the thick vegetation in this valley, he found a stream which coursed through the valley. Both parties are agreed on the existence of such a stream. Indeed Mr Ackamann went all the way to Preah Vihear to find the direction of its flow.

'Now comes a most important statement from Mr Black. The east outflow of this stream (the east outflow; that is the outflow into Cambodia) was, he says, once stopped up, so local legend had it, by the Khmer builders of Preah Vihear to turn the valley into a reservoir to provide water for the thousands of workmen engaged in building Preah Vihear. Mr Black himself saw no evidence that the valley had been used as a reservoir, but he said that the job of turning it into one would have presented no difficulty.

'In 1955 then, the east outflow of this stream was unblocked. Can one doubt that it would have presented even less difficulty thereafter than in the year AD 900 to block it, had it been desired to do so?

'Let us concentrate on the stream which Parmentier and Black have stated they saw flowing down into Cambodia at, or near, point F. Between them these two witnesses account for the presence of such a stream in 1924 (Parmentier's first visit), 1929 and 1930 (Parmentier's second and third visits) and 1955 (Black's visit). The same stream is even shown on one of the many versions of the physical features produced by Thailand's experts — their map 75c.

'To walk along a stream, to observe its course and then report what one has seen, does not require any expert knowledge at all. Anyone can do this. In particular, Messrs Parmentier and Black did do it. So when Parmentier and Black say they saw a certain stream and Mr Ackamann said he didn't, but saw a different and inconsistent stream, the three of them are speaking on the same level; that is, they are all three speaking as witnesses. Intrinsically, and particularly from the point of view of degrees of expertise, none of them is entitled to any greater weight than any of the others.

'Now, we do not assert that Mr Ackamann did not see what he says he saw. We hope and believe that the court will be impatient with assertions, if they are made, that Messrs Black and Parmentier did not see what they say they saw.'

Rejoinder of Sir Frank Soskice, Q.C. March 1962
(Counsel for the Government of Thailand)
(Former Attorney General United Kingdom)

'I will endeavour to follow the sequence in which Counsel for Cambodia addressed the Court.

'First comes Mr Acheson.'

'I wish to invite the Court's attention, particularly to Mr Acheson's statement with regard to this eye-witness evidence and the conclusions which he seeks to build upon it. It is not eye-witness evidence in any real sense but only assertions in books by persons. Henri Parmentier and John Black, whom it has not been possible to bring before the Court as Mr Ackamann was brought before the Court. Furthermore, neither of these so-called witnesses can ever have had the advantage which Mr Ackamann had of seeing the stream actually flowing. We know when they were at Phra Vihar Parmentier was there (and I take the dates as they were given by Professor Reuter at C.R., p. 548 in May and June 1924, March 1920 and January and February 1930). Black tells us in Page one of his article that he was there twice, both visits being in 1955, one was in the late spring, just before the rains, and the other was in November. Thus all the visits of these two archaeologists took place as one would expect them to take place, in the dry season. On Black's second visit to Khao Prah Vihar two tributaries of the Se Mun had to be crossed. After the rains both streams were running high and bridges were shaky. It is not correct of Sir Frank Soskice to say my second visit was made in the dry season. In point of fact, I planned by second visit immediately following the rains.

'Mr President, Members of the Court, Mr Acheson then goes on to refer, as his second so-called eye-witness, to an article by Mr John Black who visited Prah Vihar in 1955. On page 468 of the *Compte Rendu*, Mr Acheson sets out the quotation upon which he relies in Mr Black's article. Mr Black is describing the terrain just north of the broken staircase, and here is what he says:

' "The hillside falls almost abruptly into a large depression or basin before it rises again just as steeply to the rocky basalt plateau, where Nai Amphoe has so kindly built a rough sala for the traveller. But for the elephant tracks it would be difficult to wedge a way through the thick vegetation of this basin along which courses a stream. Local legend has it that this was a former reservoir. Making use of the natural depression, the builders are said to have converted it into a dam, by controlling the east outflow to provide the large water supply needed by thousands of workmen who must have been engaged in a task the magnitude of Phra Vihar. There was, however, no evidence that this natural depression had been used as such, though the job of creating a reservoir would have presented no difficulty to the Khmer who were unsurpassed in the art of water conservancy."

'Let us consider what Mr Black says. In his case, unlike that of Mr Parmentier, it is fairly clear he is referring to the valley at point F. He says that the stream runs through it. He does not say in which direction the stream runs. It is natural that he should not do so for, as we have seen, he visited Phra Vihar in the dry season, when the stream would not have been running at all. See Mr Black's statement overleaf. He does not say that he saw an east outflow. All he says is that according to local legend an east outflow was blocked at the time of building the temple, that is many centuries ago. And he is careful to add that he found no evidence to confirm this legend. There is little enough here, one might think, of relevance to the present discussion: but we shall see what Mr Acheson makes of it.

' "To walk along a stream, to observe its course and then report what one has seen does not require any expert knowledge at all. Anyone can do this. In particular, Messrs Parmentier and Black did do it. So, when Parmentier and Black say they saw a certain stream and Mr Ackamann said he didn't, but saw a different and inconsistent stream, the three of them are speaking on the same level; that is, they are all three speaking as witnesses. Intrinsically, and particularly from the point of view of degree of expertise, none of them is entitled to any greater weight than any of the others."

'For good measure, Mr President and Members of The Court, Mr Acheson proceeds to add this:

' "Now we do not assert that Mr Ackamann did not see what he says he saw. We hope and believe that The Court will be impatient with assertions, if they are made, that Messrs Black and Parmentier did not see what they say they saw."

'Mr President and Members of The Court, I gather that Mr Acheson is pained if opposing Counsel say that they are surprised at statements that he makes. In order to spare him anguish I will therefore not say that I am surprised at these statements of his; although politeness certainly requires that I equally should not be understood to say that I am surprised.'

From me! Enough of this dialogue between Messrs Acheson and Soskice about the direction of a certain stream, delineating the boundary between Cambodia and Thailand, or in other words where is the watershed on the Phra Vihar Plateau?

World Court's Judgement on the Case Concerning The Temple of Preah Vihear (Cambodia v. Thailand) 15 June 1962

1) — That the Kingdom of Thailand is under an obligation to withdraw the detachments of armed forces it has stationed since 1954 in the ruins of the Temple of Preah Vihear.

2) — That the territorial sovereignty over the Temple of Preah Vihear belongs to the Kingdom of Cambodia.

14

Search for early maps for The Siam Society

The Cartographic Section of the Society sent all the way from Bangkok to Esher in Surrey to help to build up their record of early maps from the twelfth Century onwards. While I was aware that numerous cartobibliographies could be found in Europe, this was not to be my way of getting results. Rather, I felt there existed in many cities of this continent establishments where old maps were collected and sold by experts.

At this point a word is necessary about the movement of Thai people from their origin just south of the Yangtse River in China. Yunnan Province was a mile-stone on the way south. Migration at this stage had reached a comparatively high level in agriculture and ways of marketing their products, but their aim was still south. They had an eye on the upper waters of the Chaopraya and the Mekhong where the territory was under populated and it was said 'Fish abound in the waters and rice thrives in the fields.' An expression associated with the Thai people to this day.

From the twelfth century onwards the establishment of capitals was the order of the day. Chiengmai was among the first but it is said that the Sukho-Thai Dynasty (1257-1436) was the cradle of Thai civilization. However, it was not until they reached Ayutthaya (1350-1767) further south, that this place, as a capital, will always be an outstanding one for the Thais. The present capital Bangkok (Krung Thep) was set up much later to suit Thai taste and ideals. In 1939 Siam, as the state was then named, was changed to Thailand.

As already mentioned my principal interest in retirement was in the field of marine technology which involved much travel. This made it possible for me to devote some time to my antique maps commitment. Cartography of far away places in medieval times was very much in the hands of the powerful nations: Portugal, Holland, Spain and England. In the Renaissance period Italy was spreading her wings too and for me the Biblioteca Vaticana, through their missionary connection, was a very good source of information.

After a considerable amount of groundwork ten maps were sent to Bangkok. On some of the maps certain components of the design did not form part of the information being presented, but were ancillary to it. On early maps the most prominent feature was the *cartouché*, a panel often elaborately ornamented, serving to contain title, key or dedication. This characteristic was outstanding in my gathering of maps. I have chosen two of the maps, with detail, to give the reader some idea of what was involved in this anthology of antique maps.

1. From India Office Library, London
   Map Size 37" x 28" based on:-
   D'Anville Atlas de la Chine 1737.
   The map itself was an arrangement of charts by Alexander Dalrymple, Hydrographer to the East India Company 1774-1806.
   Chart (a): Part of the River Menam in the Kingdom of Siam (Thos. Dunning Lippiatt 1797)
   Chart (b): Plan of Menam from Siam/Ayutthaya to the sea (reduced from large scale)
   Chart (c): Menam River from Kaempfer
   Chart (d): Menam River from Dutch M.S.

2. Map of India. Capital of the Royal Kingdom of Siam.
   Drawn on site by Mr Courtaulin, Apostolic Missionary to China. Sold to Paris by F. Jollain, Senior.
   Rue St Jacques a la Ville de Cologne.
   Details on the map from artistic and architectural angle: Royal Palace: Elephant Stables: Quarters, Maures, Chinese,

French, Dutch; Trading Posts.
College of Nations: Churches of St Dominique and St Paul: Grand Pagoda of the King. Plan of Lopburi and Forts of Bangkok.

Looking back over the period of my probe into antique maps I found one of interest above all others. It was a Maritime Cartogram, dated 1433 entitled 'The Mao K'un Map' concerning sea voyages from China, Champa, Siam, Java, Sumatra, The Straits, Ceylon, India, Persian Gulf, Red Sea and East Coast of Africa. Expeditions were carried out during the Ming Dynasty in China from 1368 onwards by that Grand Eunuch, Cheng Ho; (Zheng He in present-day romanization). A certain Muslim Ma Huan was his pilot. About the middle of the fifteenth century the Ming Emperors withdrew their interest in the Southern Ocean. It was into this vacuum of an ocean that the European powers, aforementioned, thought they were making a grand first entry.

15

High Flight

Returning to my technical life around the world and recalling happenings that made the journey worth while. Chris Magee, stationed in Hong Kong with US Rubber, travelled with me in many of the East Asian countries.

Once in Tokyo, when we were in adjoining hotel rooms, I was listening to a broadcast from the US Forces stationed in Japan. Suddenly my interest was absorbed by what reached my ears from the loud speaker. The subject was loud and clear — 'Outstanding poets of World War Two'! This one happened to be Pilot Officer John G. Magee Jr. An American Citizen who gave his life while with the Royal Canadian Air Force. The radio station was broadcasting one of his great poems: 'High Flight'.

When the broadcast was completed I could not get quick enough to my colleague next door with the exclamation 'Have you been listening to the US Forces radio?' His reply was simple and to the point, 'My Brother,' he said. Some years later while addressing the memoral service to those who lost their lives in the 'Challenger' disaster, when a space craft disintegrated shortly after take off, President Reagan quoted 'High Flight'.

16

Single Point Moorings in far away places

My introduction to Off-shore Engineering with special emphasis on giant tanker moorings is already part of this autobiography. The time is now ripe, however, to take a look at individual projects, in various parts of the world where events technical and otherwise were, for me, of more than passing interest.

Rome

Though not in itself a port, Rome has a road and river Mediterranean link: by road to Ostia that ancient city which takes one back to the Roman Empire. Conditions over the intervening years have removed all sea connections from Ostia. It is the river Tevere or Tiber which flows through Rome into the Mediterranean at Fiumicino, near Leonardo da Vinci Airport, which is the centre of my interest. Ostia come into the picture, but in a special way.

The port of Fiumicino provided shelter for the small craft needed to communicate and transport equipment to and from the tanker terminal a mile off-shore. The Harbour Master in charge of port operations made life easier and introduced other port users. The design of the anchorage here was known as a catenary anchor leg mooring and required a considerable area of sea-bed, firm and consistent for marine anchors designed here for a 100,000 ton vessel. This, together with flexible connections from below and above the buoy to tanker, was my job.

As well as a port for loading and discharging general cargo, Fiumicino was a fishing centre. Contact with the fishing fleet enabled us to buy the best of their catch to supplement our spaghetti diet. Closer observation on my part brought to light net failures. When I knew the fishermen better they told me they were fishing in an area where ships in Roman times approached Ostia. Shipwrecks were not unusual in these parts and a common cargo was olive oil or wine carried in amphorae. These containers, for the most part broken, were moving around on the sea bed and tended to cut the fishing nets. Occasionally a complete amphora was picked up by a net and it was my good fortune to be on the wharf when such a 'prize' arrived. The fishermen were ready to let me have the amphora, indeed only too glad. This possibly 2,000 year old antique was intact, encrusted with sea shells and had two handles on a narrow neck.

How to get the amphora out of Italy was a real problem. The 'prize' simply had to be taken by my own hand on board the plane. A soft-sided suitcase was purchased of a size to allow the minimum movement of the amphorae. At the airport booking point the official refused to accept the case as hand luggage. I was in no position to argue but finally persuaded him with a threat to use another airline. On arrival in London the customs said, 'What a prize, no problems here.'

Visits to the Eternal City were frequent. I was in a hotel close by the Vatican for some time so places associated with ancient Rome and the early days of Christianity were close at hand. And finally, just to make certain of another visit, I did not neglect to throw a coin into the Treve Fountain.

Niigata — Sea of Japan

The use of a single buoy as a bow mooring for tankers when loading or discharging oil off-shore was accepted by the oil companies as tanker size increased in the early 1960s. Niigata was one of the first ports to use a Catenary Anchor Leg Mooring and initially it was not successful due to failure of the flexible connections from the sea-bed to the buoy and on the surface from buoy to tanker. I was very much involved in the solving of this problem and from then on steel reinforced rubber hose was the answer, up to one metre in diameter and regardless of weather conditions.

In Japan there are now several such buoys capable of handling tankers of 200,000 tons dead-weight. The Japanese Steel Industry has adopted the same principle with success in the sea transport of iron ore in bulk between New Zealand and Yokohama.

In this connection I have yet to draw attention to my association with Private Legislation Procedure held in the City Chambers, Glasgow, July 1973. In essence, the Clyde Port Authority were examining a case presented by the British Steel Corporation to discharge iron-ore in bulk at Hunterston Ore Terminal, Ayrshire. But more on this later.

Ulsan — Southern Tip of Korea

One of the biggest US oil companies had a major stake developing an oil industry in the fast growing area of South Korea in the 1960s. Crude oil had to be imported and the site chosen for entry and a refinery was far from the 'madding crowd'. Here in the deep south, on the narrow strait separating Korea from Japan, I found myself with three colleagues a long way from some place to sleep at night.

An American contractor with a 'hard-case' team from California was setting-up house prior to starting work on the construction of a refinery. Obviously the refinery team were in need of the same eat and sleep accommodation as ourselves. Imagine my surprise when the 'big boss' said, 'We have no room for you lot, suggest you look elsewhere.' For miles around no such place could be found but further afield some twenty miles from our working site we came across a Buddhist monastery. After a long period of language difficulty the monks agreed to give us lodgings and do something about food. In effect we had a twenty mile journey to and from the job each day over a road of very rough terrain, and just to make us feel at home every morning, at four am we were awakened with the sounding of a massive gong.

The job off-shore took several weeks but we had no problems other than that just described. An interesting fact came to light in connection with our need to hire labour. In this respect, and somewhat to our relief and astonishment, a degree of skill was available where tools were involved which enabled us to get the job done on time. The same applied where the refinery was concerned. In the years that followed our experience there was a considerable demand in South Korea for semi-skilled workmen, particularly in shipbuilding.

Farewell to Japan, the land of sunrise, and Korea, the land of morning calm.

A Milestone Along the Way

Few go through a lifetime of three score years plus and still feel fit enough to consider: 'Where do I go from here?' It was this way with me when called to Geneva where the hierarchy of US Rubber Company had a meeting with their senior European staff. I had been in their service then some ten years and on a recent visit to New York had received a verbal promise from a senior manager that US Rubber would make it worthwhile to continue in their employ after ten years. When I raised the matter at Geneva with those concerned they were unwilling to discuss the subject because it was not in writing. There and then I decided to shake the dust of US Rubber from my feet and go it alone in the marine world of single buoy moorings for oil and slurries and model tests simulating off-shore conditions.

This involved an office at home with my wife as secretary. Then came the problem of advising those concerned what was afoot and where to find me. For over a year I had been in close contact with Imodco-Europe in Paris. They were affiliated with Imodco head-office in Los Angeles, who had their own design in mono-moorings capable of handling bulk oil off-shore. Others in many parts of the world made their own contact and I was moving around in the Middle East, North Africa, the Gulf of Mexico, Brazil and Argentina — for the most part talking and illustrating the subject on blackboard and screen.

Imodco-Europe were in shipping and had their own shipyard at La Rochelle, France, as well as access to model-testing facilities at Grenoble. My connection with Imodco, overall, was close, as the attached letter from Imodco's Chairman of the Board indicates.

Early in 1975, when exploration work in the North Sea indicated a major source of oil, the Cranfield Institute of

ARTHUR W. RADFORD
APT. S-312, 520 N STREET, S.W.
WASHINGTON, D. C. 20024

January 24, 1969

Dear Mr. Black:

At a meeting of the Board of Directors of IMODCO International, Ltd. on 15 January 1969, it was the motion of the Board to offer our sincere thanks and attendant congratulations to you for the very outstanding work that you have been performing in behalf of both IMODCO Europe and IMODCO International over the past several years since you joined our group.

Although I have not had the pleasure of personally meeting with you, I have heard much about you and the wonderful work that you have done in IMODCO's behalf. As one of the pioneers in the Mono Mooring field, it is indeed an honor to have you with our small group. And, I have been advised by both Admiral Leggett and Commander Frankel of your growing reputation throughout the world as the true expert in the ever growing field of the Single Point Mooring.

As you know, it is only by virtue of perseverance and faithful devotion in the performance of any task that success is ultimately achieved. Your devotion to the promotion of the Single Point Mooring, which has been accomplished in a truly professional manner, can -- and will -- bring, I predict, the ultimate success of the SPM concept and for the IMODCO family that we have all striven for for so long.

I am looking forward to the pleasure and privilege of meeting with you at some time in the near future. May I once again offer my warmest thanks and sincere congratulations to you.

Sincerely,

Arthur W Radford

Admiral, USN (Ret.), Chairman of
the Board, IMODCO International, Ltd.

Mr. John Black
63 "The Woodlands"
Esher, Surrey
England

Technology made it known they were planning a colloquia on the general subject of off-shore structures. I was invited to take part in this on the very wide subject of buoy mooring systems. To fall in with the requirements of the colloquia in time allowed and substance, I narrowed my subject down to off-shore and deep-water tanker terminals. In effect, the design proposed was for a single point mooring that would enable tankers to moor safely and thereafter load or discharge oil, certain chemicals and slurries. Further details are unnecessary here, except to say that a wordy exercise of this kind should have in mind selling as well as force colloquia. As it turned out later, I was to visit many out-of-the-way places world-wide to talk about and sell single point mooring terminals.

It may be, by now, readers of this story have guessed that although the main object of so much travel was connected with oil, I had in mind, given the opportunity, the following of pathways laid down by ancient civilizations. How could one go to Libya so often and not observe the antiquities of Tripolitania, see Leptis Magna, in existence long before the days of Islam and so Roman from its foundation?

In Egypt, the birthplace of modern archaeology, there is so much of interest in Cairo itself where the museum is one of the best of its kind except that there is such an overflow it would pay the authorities to display part of their wonderful history in another place. Another country which took me off oil for a while was Iran. More time is required here to get round the area involved — from the Caspian to the Indian Ocean and from Afghanistan to the Persian Gulf. But one must set aside time to see Teheran, Esfahan, Shiraz and, above all, Persepolos. The latter is the final and most eloquent expression of near east culture.

Coming to India, where time was spent along the valley of the Ganges, and on the west coast from Bombay south to Sri Lanka, from time immemorial religion was away out ahead and to this day it is a great question among archaeologists which country, India or China, had the most influence, even now, in the countries of South East Asia.

Nak'on Pathom in Thailand and Angkor Wat in Cambodia, both of whose origins are not yet understood. Finally, when it comes to China, I have dwelt at some length on how the Chinese spread their wings, given the opportunity. Suffice it to say that from my own experience, I have travelled along the Old Silk Road and into the deserts of Gobi and Taklamakan in search of oil. There I came across The Caves of The Thousand Buddhas where in one place there is to be seen the best and most voluminous presentation of Buddhist art in existence. In these caves there is an element of Greek art to be seen. But on my other side, close by The Caves of The Thousand Buddhas, an oil discovery was made where the Great Wall of China meets the foothills of the Tibetan Alps.

17

Political and Legislation

The Clyde Port Authority
(Hunterston Ore Terminal)
Private Legislation Procedure
Scotland Act 1936

This inquiry, both political and legal was held in Glasgow City Chambers 18 July 1973. It should be explained why the action took place in Glasgow and how it involved the writer.

The handling of iron-ore from the British Steel Works near Motherwell in Central Scotland was the reason for the inquiry presided over by Viscount Stonehaven, A.I. Struct., D.L. and other politicians from the House of Lords and Commons. The port of entry was Hunterston, Ayrshire. The ships alongside the wharf discharged their cargo (iron-ore) close by in mountainous heaps, awaiting rail transport in open trucks, to Motherwell. One can imagine the attitude of near-by residents, living in an atmosphere of iron-ore dust.

The 'Promoters' were in favour of 'no change' in the existing method of discharge. On the other hand, the North Ayrshire 'Objectors' proposed another method of unloading the iron-ore in slurry form, through a flexible pipeline from the ship moored off-shore, then, once ashore, by steel pipeline to Motherwell, still in slurry form, thereby assuring a dust-free atmosphere throughout. I was one of four technical experts acting on behalf of the 'objectors'. The 'promoters', however, succeeded, through pressure from British Steel Corporation and The Clyde Pilots

Association, in preserving the status quo.

A final word to complete my view point. As a technical adviser, shortly before this inquiry in Glasgow, I was involved in solving a similar transfer of iron-ore from New Zealand to Japan where a single point mooring was in use at both terminals and the iron-ore was slurried at point of entry in Japan. Needless to say, it was a successful dust-free operation.

**A case of Arbitration
between
Norwegian Owners of Tanker** *M/S Sidney Spiro*
**and
Chevron Tankship (UK) Limited, London
Charterer**

One more legal case involving an Arbitration Decision and a Demurrage Award.

A Single Buoy Mooring at Canaport in the Bay of Fundy, some 4,000 feet off-shore had an accident with 'Selflote' hoses connected to the buoy. The delay in repairing or replacing the hose was the cause of demurrage to the tanker *Sidney Spiro*.

I was introduced to the case through an attorney in London. This was followed by a visit to Canaport in the Bay of Fundy for inspection. A final decision on demurrage and who was responsible was made in New York. My evidence was given there under: 'The facts and contentions of the parties' as an expert having extensive experience with single buoy moorings and related hose technology. The decision and award need not be gone into in detail here. Suffice it to say, it appeared to the writer a 'Solomon's Judgement'.

An outstanding feature of this assignment was my hotel, 'The Algonquin', the only hotel of its kind in New York — or anywhere else.

18

A Chinese Connection

This episode takes me back to my Shell days in Shanghai when I had a responsibility in the Marine Department. The senior Chinese in this section was Clement Chen. He was painstaking and likeable, did a good job and was familiar with the many problems associated with transporting oil 2,000 miles along the Yangtse River. In 1949, when the Communists ruled the roost in Shanghai, he escaped to Hong Kong with his family. One day he called to see me at Shell to ask if I could help in signing US Consular forms to get his son, also a Clement, a passport to the USA. I was very willing and soon after the boy left for America. In his late teens he succeed in graduating to allow university entrance. He studied architecture and when he crossed my path again, a good few years later, he was a director in Holiday Inns and their chief architect.

How did I meet him again? He was in London with his wife, June. And he made contact, visiting us in Esher. He was anxious to be friendly with his father's old boss and was ready to do anything, even as far back as following the teaching of Confucius to do a good turn to what he regarded as a family connection. Early in 1984 he found a way.

We received a letter from Clement Chen which said:
'This is the Year of the Olympics in Los Angeles and I invite you to be my guest at the Games. Not only me but my wife, two daughters and their families.' Accommodation was reserved for us in the Holiday Inn, Pasadena. Chen was the owner. He had tickets for all sports at the Games and we saw many of the

world's best athletes. Transport by car was always arranged, and on one notable occasion he chartered a plane and took us all to the Grand Canyon, for several days.

The home of the Chens in the capital city of California must be mentioned and Clement's impressive office in Montgomery Street, not forgetting the many touching sights of San Francisco. The Holiday Inn at Palo Alto was our abode while in residence.

The Chen family came to the airport to wish us *bon voyage* and farewell. Sadly, Clement died a year ago when in the middle of building a hotel in Sian (Zian), his homeland. The hotel is named Jianguo and has another of the same name in Peking (Beijing).

What a man, with such a philosophy of life!

FINALE

This retrospect has now reached the winding-up stage. In 1975 work and travel are almost over. There is, of course, always an exception, for Rena and I it is Switzerland, the home of our eldest daughter Fiona and two grandsons, Mark and Simon. June and John in the latter's diplomatic post have added to their experience in Tel Aviv, Bangkok, Ankara and Kuala Lumpur. Sadly, a few years ago Sam in Switzerland, Fiona's husband, died suddenly and recently John, June's husband also. Both men were in their fifties.

This short prayer has been in my mind ever since:

> 'Father, Lord of Heaven and Earth and Father of our Lord and Saviour Jesus. Sam is on the way and John follows him to be with you. Their minds are now overflowing with the joy they will experience when they see you face to face.'

Rena and I are in Cresuz, Switzerland, once a year. Our stay used to be at snow time for skiing, but as the years go by we prefer the summer. Fiona's chalet is a gem, it is 1,000 metres above sea-level and the view is beyond compare. The boys are her mainstay and do well in handling a very difficult hill-side garden. Mark is now seriously studying in a college for hotel management in Lausanne. Simon, still at school for another year, is now thinking of a future occupation, possibly graphic design.

From time to time in the course of these memoirs through all the countries, China, Korea, Japan, South East Asia, The Persian Gulf and North Africa, I have drawn attention to outstanding manifestations of human cultural and intellectual achievement. My job came first but I never failed to follow, given the time and opportunity, the pathways from ancient times which ultimately led to our modern world. In this connection the Royal Geographical Society and the Royal Scottish Geographical Society awarded fellowships.

LECTURE AT CENTRE FOR FAR EASTERN STUDIES LONDON UNIVERSITY
26 November 1974
A Kan-Su — Ching-Hai Journey (NW China)

The Old Silk Road from Lan-Chow to Tun-huang and The Caves of The Thousand Buddhas Ch'ien-fo Tung.

From Lan-chow to Si-ning, the Lamasery of Kum Bum (T'a-erh Ssu) and the Koko Nor (Ching-hai Hu)

Introduction:

The entire terrain, over mountain and desert, was surveyed from the air. Following this, a journey of some 1,800 miles, was undertaken along the old caravan routes by truck, camel, mule carts and often on foot.

Unchanging Features:

The great expanse of mountain and desert along the Northern edge of the Tibetan plateau and on the fringe between, the interesting oasis townships of antiquity.
 In the east of the region, the fantastic loess mountain desert

country with the Yellow River and its tributaries.

The highest mountain is some 20,000 feet and the average valley 3,000 feet while the majority of habitations lie between 4,500 and 9,000 feet above sea level.

In some mountain districts there may be as little as 100 no-frost days in the year. In the Kan-su Corridor and west, the rainfall is low and the atmosphere dry. Overall considerable drop in temperature and wide variation in weather conditions can be experienced in twenty-four hours.

Lan-chow to the pass where the Wu-sha-Ling and Nan-shan meet

Lan-chow, the provincial capital, is impressive from the air, with the moving mass of Yellow River honeycombing the city and, in the immediate surroundings the great folds of loess desert mountains. Within the walled town itself the newcomer cannot fail to be surprised at the Arabic script over many shops. Although the sons of Han still predominate, from here on westward, an admixture of peoples are encountered, the Tungan or Hui Hui and, of course, the Mongol, not forgetting the Tibetan or Tangut who left a mark on Kan-su when his Hsi-hsia kingdom reigned supreme about 800 years ago. Divide and rule, an attribute to Britain's Imperial past, has long been and probably still is, practised by the Chinese in these parts.

At Ho-kou — thirty miles west of Lan-chow — three rivers meet, the Hwang Ho, Sining Ho and Chuang-lang Ho. With nature's gift of such a confluence of waters, coupled with the head of pressure from the latter, coming from the eastern slope of the Nan-shan, the Chinese use their flair for hydraulics to harness the water for power and irrigation. From Ho-kou to Yung-teng, at half-mile intervals, in about thirty miles, canals off the Chuang-lang Ho provide power through vane-wheels for grain milling. To supply the irrigation channels, the traditional peripheral pot wheel takes over. Only here, in the Lan-chow area, the pot wheels come in giant-size, some are over forty feet in diameter. Although the wheels do not look strong, creaking as they turn, nevertheless, they are stoutly built to withstand strong river currents. Carriage by water is always cheaper than by land and here the sheepskin raft comes into its own for the transportation of both people and goods. Grain cultivation is extensive and fruit growing is a feature, thanks to the rich loess soil.

The loess deposit can pack itself in depths of 100 feet and more. It is highly erodable to rain, wind and river action, giants like the Yellow River have the appearance of solid mud on the move. The farmers are presented with a real problem, in holding the loess to prevent erosion and thereby hold the temperature and water in the ground. To do this, the fields are covered with a gravel mulch. This layer of gravel — each stone about the size of a fist — completely covers the soil and has the desired effect, but the problem then presented to the farmer has to be seen to be appreciated. He must employ a ploughing technique to keep the gravel on top. Fields of grain, cotton and vegetables are grown extensively under these conditions.

Climbing away from the loess country along the Old Silk Road, now known as the North West Highway, the Mao-tao-shan Pass lies ahead. At 9,000 feet, this pass between the Nan-shan and Wu-shan Ling mountain ranges is the great divide, the dust of the loess country is behind and the great desert of Gobi ahead. Although geographers do not show it as such, in reality it is the entry into Central Asia.

The Mao-tao shan Pass to Tun-huang

Wu-Wei, the first town of importance on the west side of the pass, is still in the green belt and is often referred to as the granary of Kan-su.

The Great Wall has been encountered in various stages of decomposition since leaving Lan-chow. Yung-Chang, the next halt 6,000 feet above sea-level, is under the shadow of the Wall. The hotel was drab and after three hours of travel in chilling rain with sodden clothes, the place was indeed cheerless. Next morning the local officials seemed determined to make up for a miserable night by arranging a breakfast in the form of a full-scale Chinese feast with all the trimmings.

Arrangements were made after breakfast, for a journey of 150 li by mule cart into the neighbouring province of Ning-hsia, the first taste of Gobi the hard way, on the ground. It is perhaps as well to say now that the word 'Gobi' is not the proper name of a geographical area but a common expression used by the Mongols to designate a definite order of geographical features. These are wide shallow basins of which the smooth rocky bottom is filled with sand, pebbles or, more often, gravel.

Passing through the Great Wall, the first halt was at the Gobi village of Ning-yuan Pu. The village was served by a clear stream — a rare thing in this part of Kan-su and which later disappeared in the desert. After the discomfort of the previous night at Yung Chang, the school here was luxurious as a rest house. Ning-yuan Pu was an excellent place to stock up with vegetables, eggs and chicken for a week in the desert.

Water holes in Gobi can be dry, so at the last outpost Hsi Po the stock was brought up to capacity. One, Sun, by name, was the owner of the outpost, he was a farmer, obviously well-off with arable land and a water supply in the region of what can only be described as the Sun-farm fort. The inner sanctum some seventy-five yards square was laid out in Chinese style with a centre courtyard. This was protected by mud walls forty feet high and tapering in thickness from twenty feet at the base to twelve feet at the top. Four square towers, set into the wall at the corners, were for the dual purpose of storing grain and providing a look-out. Large rocks were also stored there for defence. A specially constructed shute guides the rock on its way to injure if not to crush the undesirable who dares pass through the outer wall which in itself is a substantial barrier. Hsi-Po was a trading outpost. The settled farmer with agricultural products exchanged them with roaming bands of Mongols for skins, camel hair and wool, the currency of the nomad.

The mule cart was exchanged here for the camel. A long camel ride over the gravel of Gobi is a never-to-be-forgotten experience. The bactrian camel is quite a different animal from the dromedary of the Middle East. Indeed, a number still exist in their wild state but it is said of them they are the dodos of tomorrow. No saddle is used on these Central Asian camels and added to a frightful gait, is a bad temper which comes quickly to the surface if his load, rider, plus freight is over 300 pounds. The result can be a shower of regurgitated cud on the rider — a nasty experience.

Back on the Old Silk Road, travelling by truck and journeying west, with the Tibetan Alps on the left and Gobi to the right, a succession of oasis towns lie ahead where the rainfall is sparse and much of the earth is saturated with salt and soda. The next halt at Shan-tan is right in the corridor or panhandle of the province. The cultivable area in this oasis has a good system of irrigation supplied by springs. Shan-tan was a town of temples, now, for the most part, used as dwelling places. In its heyday it was a key-point in the Old Tangut Kingdom of Hsi-Hsia and offered much resistance before capture by Genghis Khan. The Khan, a pastmaster at the art of pillage and destruction, razed everything to the ground but the temples. As a result of this, the Buddha and Kuan Yin now grace the interior of many a living room and even workshop.

The corridor town of Kan-chow is the most extensive oasis, having a good supply of water. Highlands to the north trap and throw back fast-flowing torrents, preventing valuable water from the glacial streams of the Nan-shan disappearing in the Gobi.

Between Kan-chow and Chiu-chuan, a distance of 120 miles, there are ruins of ancient townships long since deserted, possibly due to lack of water and over many years' warfare, rebellions and indifference of central and provincial authorities.

The next oasis town, Chiu-chuan, on the eastern edge of the Chiu-chuan — Yu-men basin, roughly an oval in shape, some eighty by forty miles, depends for its water on the melting snow. Here water rights are more fundamental than land rights and every able-bodied man and boy must contribute labour to ditch and canal the water fanning out from the Nan-shan foothills before entitlement can be claimed to breach the canal close to the family fields. By dint of very hard work in this barren area, farmers are able to make a living growing wheat, vegetables and fruit.

It would appear to be something of a misnomer to refer to this area as a basin, with the highest point at 10,000 feet on the alpine slopes of Northern Tibet and the lowest at 2,000 feet elevation on the fringe of Gobi. In point of fact, the principal object of this journey was to examine the geology of the Chiu-chuan-Yu-men basin, so a month was spent in this area to get acquainted. Apart from the survey there appeared at first to be little of interest in this barren land. But after a month's stay it turned out to be fascinating for a number of reasons.

The Great Wall which had been a companion now for over 400 miles from Lan-chow, terminates in the foothills of the Tibetan Alps ten miles west of Chiu-chuan. Kia-yu Kuan, the last fort on the wall, is well-preserved, as are long stretches on either side of the fort but as the barrier terminates at its western extremity, five miles from Kai-yu Kuan, it is again fragmentary.

From Shan-hai-Kuan, the eastern extremity of the barrier on the Yellow Sea. The wall makes its away across the mountain, valley and desert for roughly 2,500 miles. It can now be said it is the only man-made object that can be seen from the moon, but it has long been known, for another record, as the longest cemetery in the world.

Another sight, as old as antiquity, worth recording in the Chiu-chuan Yu-men basin, was revealed by large seeps of oil at the surface outcrops of fault planes. The oil is skimmed off the streams and has been used since ancient times in lamps and to grease the large wheels of desert carts.

From certain vantage points, though few in number, that magnificent creature, the Ovis Poli, can be seen with the naked eye, jumping from crag to crag. Perhaps not so spectacular, but with an attraction all of its own, herds of antelope were seen in an area of the foothills where, in the dim and distant past, there was once a great lake.

Little has been said about food and goods shown in the shops. For the most part it was aimed to live on the land, carrying the minimum of canned goods. Many of the inns were very rough and ready and the highlight on the menu was mien flavoured with cabbage water. In the 1947/48 period there was great poverty in Western Kan-su. Compared with the Yangtse valley or even Lan-chow, the goods in the shops were of poor quality, suggesting a much lower standard of living.

The journey from Yu-men to An-si brought into relief a great stretch of desert known as the Black Gobi. Here the gravel is of pre-Cambrian age and fossils are found, including a number of sea shells.

An-si was a ghost of its former self. The countryside, where once prosperous townships existed, has gone back to the desert through lack of water. Judging from maps made by Aurel Stein when he travelled in these parts over forty years ago, there has been a considerable deterioration in what was once a fertile area. In this respect, the magistrates of An-si pointed out that the population has fallen from 100,000 at the beginning of the century to 3,000 in 1948.

In the intervening period, glacial streams from the Nan-shan have disappeared below a conglomerate of gravels cast off from the high mountains. In fact, throughout the length of the Northern Tibetan Escarpment lies a basis of piedmont gravel covering in parts a width of forty miles or more and everywhere utterly barren. This is aggravated in Western Kan-su by a very active uplift along the Nan-shan and Altryn Tagh Ranges, where earthquakes are common. The result is destruction of the irrigation systems on the piedmont outwash north of these mountains.

While on a short reconnaissance in Gobi close to An-si, a band of some forty Kazakhs — including families — were encountered. Their tents, six in number, were about ten feet in diameter with four feet six inch walls, then tapering cone-shaped to an overall height of eight feet. In terms of this world's goods they were rich, with over 600 in their herd of sheep, goats, horned cattle, horses and camels. They were pastoral nomads sharing a common history with the Kazakhs of the USSR and belonging to the Turkic group. Regarded locally as notorious raiders, on the contrary, it was found the Kazakhs were very friendly. It was amusing and interesting to watch their women making ropes from animal hair, using fingers and toes with great dexterity. Information was obtained later from the Garrison Commander of An-si that in the area between Tun-hung and An-si about 1,000 Kazakhs were on the move eastwards. This movement, without doubt, was dictated by politics. An estimate of the number of Kazakhs in Hsin-kiang province is 300,000.

Tun-huang, a much bigger oasis than Yu-men or An-si, is also a shadow of its former self. Fruit was plentiful and there was a good acreage of grain and cotton in the area. The melons came from the next province Hsin-king where the famous Hami Kua (melon) from the Turfan Depression has no equal.

On the edge of the Lop Nor marshes is Tun-huang, which takes its name from Han times when it occupied a key position as a commercial centre, where it was the meeting place of two important highways. The road from the Chinese capital Si-an and the road from the Mongolias joined at Tun-haung. From this focal point two routes carried the traffic to and from the west, one north of the Tibetan-Tsaidam plateau along the foothills of the Altyn Tagh and Kun-lun, the other through the Tarim basin, north of the Taklamakan Desert south of the Tien-shan — this was known to the Chinese as the Tien-Shan

Nan-Lu. The name 'Great Silk Road' is western terminology and is connected with both routes which become one at the oasis of Kashgar, a thousand miles to the west of Tun-huang.

In the ten miles from Tun-huang to The Caves of The Thousand Buddhas, a battle can be clearly seen between two giants — Gobi and Nan-shan. The desert stretching its long arm of sand tends to blanket the foothills and, as it was seen then, appears victorious over the gravel-shedding mountains. By and large, there is little fasciation about great tracts of sand, but after a while they fit into a picture. Hereabouts people talk about the singing sands of Gobi and the desert stones are a revelation in both shape and colour. Triangular stones point like arrowheads in the direction of the prevailing wind and show the scouring action of the sand on two sides. Leaving Tun-huang, a few days later, on return east, a severe sand storm was encountered, strong enough to scorch the skin and be a reminder of the violence and persistence of the Gobi winds.

BUDDHIST MONASTIC CENTRE — CH'IEN-FO TUNG

THE THOUSAND BUDDHA GROTTOS NEAR TUN-HUANG

The conglomerate cliff from which the caves are cut was laid down in the first place by torrential streams from Ch'i-lien Shan — a range in the northern escarpment of the Tibetan plateau. At a later stage high gradient streams cut minor narrow valleys from the outwash of gravels which in turn were cemented with a soft white calcite. The result was a cliff ready-made for easy excavation.

The caves are cut along the cliff face, in tiers, for a distance of over a mile. A panorama outline of the cave front is attached to these notes with something of a description of the grottos, their history and development.

Much has been recorded about this famous Buddhist monastic centre, but the shrines have to be seen to be fully appreciated. They are distinguished for many reasons.

As a gallery of wall paintings and statues, it is the most extensive in the world covering the first millenium of our era. The inspiration is basically Buddhist art in its imagery, symbolism and ideals of form but landscapes and scenes depicting the life of the times in India and China are prominent. Central Asia played a tremendous role in the dissemination of other cultures which influenced art at the Ch'ien-fo Tung. Indeed the Kushan Empire embracing parts of India, Iran, China and USSR (BC 2nd/1st century to AD end of 4th early 5th century) was the centre from which Buddhist culture spread across the face of Asia. The importance of this Empire lay in its geographical position straddling the trade routes of Rome, Iran and China and in its religious tolerance and coexistence of the traditions of various peoples, it was a model. This influence survived long after the downfall of the Kushan Empire.

Perhaps the most important fusion of all was the introduction of the Chinese art style from the middle kingdom itself. This illuminating element in the paintings, together with numerous Sanskrit texts brought from India by the Pilgrim Hsuan Tsang and coupled with the Central Asian influence provides a background for the study of Mahayana Buddhism. As a result, an entirely new phase in Buddhist art emerges at the Ch'ien-fo Tung. The art of China during the Tang Dynasty was at its grandest. The best period was from the sixth to eighth century, indeed were it possible to restore masterpieces of Tang artists for examination now, it is probable that the age mentioned would rank with the greatest period of creative art in the world's history. The shrines at the Ch'ien-fo Tung portray the most authentic Tang landscapes.

These shrines were the resting place for learned Chinese, Central Asian and Indian monks in their extraordinary landward journeys throughout the Buddhist world in the mid-period of the first millenium AD. In 629 AD, the most celebrated pilgrim, Hsuan Tsang, travelled by the northern trail of The Old Silk Road, via the Tarim Basin, known to the Chinese as the Tien Shan nan Lu. He returned in 644 by the southern route of The Old Silk Road via Yarkand, Khotan, Lop Nor and Tun-huang.

Tibetan predominance at Ch'ien-fo Tung was at its height from the middle of the eighth to well into the ninth century and again in the Hsi-Hsia — Tangut period when Tibetan Buddhism with its Tantric divinities found a place in the art of the caves. A rubbing of the Buddhist Votive Tablet, presented to the cave temple Chüeh by name, was a personal gift while visiting the caves. Known as the Mo-Kao Chüeh Stele, it is a famous monument because of its inscription in six forms of writing. The inscription itself is the well-known Tibetan Buddhist incantation — Om, Mani Padme, Hum — 'Oh the Jewel in the Lotus'. The forms of writing are illuminating in that they provide a background to some of the languages used at the Ch'ien-fo Tung.

> SANSKRIT, TIBETAN, CHINESE: HSI-hsia (Tangut) — rare script of the state of Hsi-hsia which occupied part of Kansu, including the Tun-huang region and Ning-Hsia.
>
> MONGOLIAN (Phagspa) and UIGHUR: Runic writing of Central Asian Turks.

The stele is dated AD 1348, the eighth year of the Chih-Cheng period of the Yuan Dynasty. At the foot of the stele the names of the donors, including a prince, are engraved. They appear to be foreign names transliterated into Chinese.

NOTE: The Ch'ien-fo Tung owes its greatness as a cultural centre to its benefactors — merchants, army commanders, religious societies, royal and princely donors. In point of fact, in most caves, the bottom register of the walls is reserved for portraits of donors.

The Aurel Stein collection from the Ch'ien-fo Tung in The British Museum contains the Sanskrit Buddhist work of the Diamond Sutra in Chinese translation. Printed in AD 868, it is the earliest dated specimen of block printing. Other unrecorded works of Buddhist scriptures in Sanskrit are in the Museum together with Taoist and Chinese classical texts and unknown Nestorian Christian and Manichaean records. The Stein collection also includes remains of silk brocade, terracotta and stucco figurines preserved in a dry atmosphere and embedded in Gobi sand for more than a thousand years.

Commanding Watch Stations

In the second century BC, the Han Emperor Wu Ti built a chain of watch stations as fortifications in the shape of towers beyond what was later to be determined as the western extremity of the Great Wall at Kia-yu Kuan. These fortifications are not, as far as can be observed, an extension of the Wall which was erected for defence to protect China proper from the incursions of the Hsiung-nu. The forts were built to establish Chinese military presence in the Tarim Basin, protecting their commercial interests. It was in the vicinity of one of these forts that a sandstorm of considerable severity was encountered. The violent and persistent winds of Gobi can pick up the sand which, when really wind-borne, can scorch the skin.

Lan-chow — Si-ning — Koko-nor

In the upper Si-ning river valley the loess mountains taper off to give place to still higher mountains, green by comparison, but still having a thin spread of loess. The farmers have terraced the mountains hereabouts to an elevation of 7,000 feet.

In the wide open spaces of Ching-hia, agriculture takes second place to the rearing of cattle. The Yak comes first in order of importance, followed by the cow, horse, donkey, mule, sheep, goat and a small number of camels. It was interesting to see the black, as distinct from the white, Yak in the area. The former is reared in the south and west and statistics show they number 180,000. The white Yaks are from the north and east and number 20,000. The welfare of animals in Ching-hai plays a considerable part in the economy of the province. Unfortunately, there was a lack of specialists in animal husbandry, so foot and mouth disease was prevalent and, to a lesser extent, but still a real problem, rinderpest and pneumonia.

There was ample evidence of progress in Ching-hai. The roads were well-constructed and gangs were at work blasting out rock to make river gorge passage easier. Approaching Si-ning, the road surface improved and irrigation ditches were covered over. Open ditches were a constant source of trouble in Kan-su. Planned afforestation was evident. The planting lay-out was calculated, not only to prevent erosion, but in the long term to make the Huang Shui (Si-ning River) a clear stream. A feature of planned agriculture where terracing is employed was to plough around the contour of the mountains to conserve valuable rainwater. The period between crop planting is designed to reduce wind erosion. The problem of defeating loess erosion by holding the shifting dust is of major proportions, calling not only on ingenuity but very hard work.

The Governor of the province, with the capital at Si-ning, was a Muslim. It will be recalled that Si-ning was the centre of the Mohammedan Rebellion at the end of the last century, when there was considerable animosity between Han and Moslem Chinese. Ma-pu Fong, the Muslim Governor, in his late forties, laid out the 'red carpet' and even went to the extent of presenting fur coats of black unborn lamb skin. Unfortunately, when the coats reached the warmer clime of Shang-hai, lack of treatment in the early stages forced immediate disposal. Whatever might be said of the Ma family, and in the NW there were a number around in important posts, they proved themselves good administrators.

From time immemorial, Si-ning has been the junction point of two of the most important caravan routes connecting Western China with southern Mongolia toward the northeast, and eastern Tibet toward the southwest. In its vicinity too is the celebrated Buddhist monastery of Kumbum — known to the Chinese as T'a-erh Shih — which has made this region one of capital importance for the religious and artistic history of medieval Buddhism. The display within and outside the Lamasery was magnificent, with prayer wheels, tapestries and the golden finish on statues, motives and temple roofs. The Hall of Classics, with its forest of bright curtains and massive central image, surrounded by huge butter lamps, can accommodate 2,000 lamas, all sitting akimbo on cushions. On an outside porch pilgrims were falling down on their faces with outstretched arms in front. This act of homage with its hand action on the smooth wood of the porch has over the ages worn deep grooves in the floor planks. Pilgrims were arriving, in a continual stream, with tribute to keep the idle lamas, who appear and smell to be the sworn enemies of cleanliness, doing nothing all day but yawn and tell their beads.

Kumbum's greatest claim to fame is that it was built on the site of the birthplace of Tsong Kaba, the celebrated reformer of Tibetan Buddhism in the fourteenth century. Tsong Kaba founded a sect that had a dominating influence in the Tantric worship of Gautama's ancient creed common in Tibet and the Himalayan countries.

Another claim to fame at the time of visiting Kumbum was that the Panchen Lama and his satellites resided there. Passing through a moon-gateway guarded by two provincial soldiers, a further armed escort was provided to the audience chamber. Yellow was the predominant colour, though red was also very much in evidence. The visitors sat in two rows facing each other, with a centre passage between. The Panchen Lama when seated was able to look directly at his foreign audience of five while to see him, heads had to half-turn in his direction. The boy Lama was then twelve years of age and was discovered as a reincarnation of the former Panchen at the age of five. He looked bright and intelligent and was no doubt being trained by his three attendant tutors, who constantly appeared to have their heads together, so that in time he would prove as capable and crafty as his predecessor. Before being introduced, a member of the party stood before the Lama, bowed and handed him a long strip of silk. This was later returned as a memento of the visit. Pictures were then taken of the group and the Buddha, then finally each member of the party took a picture of the Panchen Lama.

In contrast to the pomp and splendour of Kumbum a few miles away the purely Tibetan village of Luksar was squalor indeed. The quaint dress and headgear of the people was most interesting but their close association with the Yak, their beast of burden, and the high smell coming from both suggested they surely must live together.

En route to the big lake Koko-Nor from Si-ning, the valleys were well-cultivated. Some twenty-five miles from Si-ning a loop of the Great Wall was encountered. This is regarded as the ethnographical border which formerly demarcated China from Tibet. A fascinating expedition awaits the traveller who would explore the course of the old frontier along the trace of the Wall where apart from a small Han (Chinese) element, the country is peopled by Moslems, Tibetans and Mongolians.

The lake is first sighted about twenty miles distant from a ridge near the summit of the Sun-Moon Mountain, 11,000 feet above sea-level. From this strategic approach to the lake can be seen three old walled towns now in a state of collapse. Relics found in and around these ancient towns indicate their Han origin. In this region, with access to China through the valleys of the Yellow River and its tributaries, the towns were used as garrison posts to keep the barbarian nomads from penetrating beyond the borderlands of Tibet.

About eighty miles from Si-ning surrounded by rolling grasslands lies the Koko-Nor, named by the Mongols who first inhabited the region — the Chingai-hai Hu to the Chinese or Clear Lake and the Anum Bor — west sea — to the Tibetans.

The Lake is 10,250 feet above sea-level, some 300 feet deep and around its shores the distance is about 240 miles. Hai-hsin Shan, the largest island of five, has a Lamasery, but the lamas leave the island only in winter when the lake is frozen. The Koko-Nor forms a giant basin with a vast drainage area from which more than fifty rivers and streams pour their contents. There is no outlet but it is thought at one time the lake drained into the Yellow River. The water is brackish and certain fish seem to like it. The encircling mountains are about 16,000 feet. Rolling grassland surrounds the lake and stretches away towards the marshes. It is wonderful country for grazing herds and considerable advantage is taken of this pursuit. Thousands of cattle could be seen and many black spidery yurts or tents, protected by ferocious dogs, were much in evidence.

South and East of Lan-chow, about ten miles off the main highway to Si-an is Shing-lung Shan the burial place of Genghis Khan — so far as is known for the second time. When the Japanese occupied Inner Mongolia in the late 1930s, the Chinese removed the silver casket containing the remains of the Great Khan and one of his wives. This was done to prevent them from using the casket to rally the Mongols to their side. At Chinese insistence — a very astute move — the remains were accompanied by living descendants of the Khan who settled at Shih-lung Shan. It is a place of rare beauty in a cradle of the mountains. The adjoining hills are green and thickly wooded and the shrubs and wild flowers were resplendent during the visit. The king of the mountainside is the spruce, but the oak, aspen, mountain ash and silver birch grace the slopes making the scenery, with the rising mountain range behind, beautiful indeed. It is said the casket has since been removed, two places are named — the Lamasery of Kumbum near Si-ning and the other to a mausoleum in the Ordos area, a place set aside by the People's Government as the cradle of the revolution.

<u>SITUATION</u> THE CAVES ARE LOCATED IN THE DESOLATE BORDER COUNTRY WHERE GOBI & NANSHAN MEET CLOSE TO THE KANSU-SINKIANG PROVINCIAL BOUNDARY.
750 MILES WEST OF LANCHOW THE PROVINCIAL CAPITAL OF KANSU. IN THE HSIEN OF TUNHUANG 12 MILES S.E. OF TUNHUANG ITSELF. THIS ONE TIME PROSPEROUS CITY IS ON THE OLD TRANS-EURASIAN SILK ROUTE.

CAVES OF THE THOUSAND BUDDHAS
(CH'IEN FU TUNG)
千 佛 洞
TUNHUANG OASIS — KANSU

HISTORY THE FIRST CAVE TEMPLE WAS EXCAVATED IN THE YEAR 366 A.D. IN THE SECOND YEAR OF CHIEN YUAN (PU CHIN DYNASTY). THE CONSTRUCTION PERIOD EXTENDED THROUGH THE DYNASTIES OF NORTHERN WEI, WESTERN WEI, SUEI, TANG & THE FIVE DYNASTIES DOWN TO SUNG AND YUAN.
THEIR VALUE LIES IN THE PICTURE OF THE TIMES THEY PRESENT, SHOWING AS THEY DO THE HIGHLY CIVILISED STATE OF CHINA DURING THE FIRST 800 YEARS OF THE CHRISTIAN ERA.
IN ART, LITERATURE, RELIGION AND SOCIAL CONDITIONS THE RECORD SEALED UP IN THESE GROTTOES FOR CENTURIES BY THE GOBI SAND IS PERHAPS THE MOST REMARKABLE ARCHAEOLOGICAL DISCOVERY OF THIS CENTURY.

ARTISTIC MERIT — THE STATUARY IN STUCCO AND WALL PAINTINGS FOR THE MOST PART PORTRAY BUDDHIST ART. THE PLENTIFUL REMAINS ATTEST TO THE SCULPTURAL TRADITIONS WHICH GRAECO-BUDDHIST ART HAD DEVELOPED AND CENTRAL-ASIAN BUDDHISM TRANSMITTED TO THE FAR EAST.

FROM THE POINT OF VIEW OF ARTISTIC MERIT THE SCULPTURE AND WALL PAINTINGS MAY BE CLASSIFIED AS FOLLOWS:- THE ART OF THE TWO WEI DYNASTIES POSSESS THE ROUGHNESS OF A PRIMATIVE STYLE. HERE INDIAN ART IS APPARENT, THE COLOURS ARE EXCELLENT AND THE LINES, THOUGH SIMPLE, ARE OFTEN CHARACTERISTIC OF THE IMAGINATION. A HIGHER STANDARD COMES TO LIGHT IN THE ART OF THE SUEI AND TANG DYNASTIES. THE TANG PERIOD SAW THE PEAK

OF PROSPERITY, ARTISTIC EFFORT AND ACTIVITY IN CONSTRUCTION OF THESE SACRED CAVE SHRINES. THE RICH COLOURS & THE REALISTIC STUCCO SCULPTURES OF THE TANG TIMES SHOW A HIGHLY DEVELOPED ARTISTIC SENSE.

THIS EXPRESSIVE POWER SO RICH IN THE TANG DYNASTY IS NOT REPEATED. IN THE FIVE DYNASTIES THAT FOLLOWED, AS WELL AS THE SUNG AND YUAN, THE DULLNESS AND SIMPLICITY OF THE COLOURS, LACK OF IMAGINATION & REPETITION OF THE SAME FORM, IN SCULPTURE, TESTFY TO THE DEGENERATION OF ARTISTIC EFFORT IN THIS PERIOD.

EXTERIOR DESCRIPTION THE CAVES ARE CARVED INTO THE PRECIPITOUS CONGLOMERATE CLIFFS OVERLOOKING FROM THE WEST THE MOUTH OF A BARREN VALLEY. A SMALL STREAM DESCENDING FROM THE WESTERN MOST PORTION OF THE NAN-SHAN RANGE HAS CUT ITS WAY THROUGH THE FOOTHILLS OVERLAIN BY HUGE RIDGES OF DRIFT SAND. THIS STREAM KNOWN TO THE CHINESE AS YEN CHUAN (岩泉) OR ROCK SPRING HAS MADE IT POSSIBLE TO IRRIGATE THE FRONTAL CAVE SECTION AND HAS TRANSFORMED THIS DESERT AREA INTO AN ATTRACTIVE OASIS. HERE POPLAR TREES ARE NUMEROUS AND ENOUGH FRUIT AND GRAIN IS GROWN TO MEET THE HUMBLE NEEDS OF THOSE WHO TEND THE CAVES. THE CLIFF FACE FOR A LENGTH OF OVER HALF A MILE HAS AS MANY AS 1000 CAVES,

LARGE AND SMALL, CUT OUT OF THE GRAVEL
CLIFF. THEY HONEYCOMB IN IRREGULAR TIERS
ALONG THE SOMBRE ROCK FACES OF THE PRECIPICE,
ONLY THE IMMOVABLE WALL PAINTINGS AND
SCULPTURES IN 425 CAVES REMAIN. IN THESE
CAVES A TOTAL OF 15 MILES OF WALL PAINTINGS
AND OVER 2000 STUCCO STATUES REPRESENT
THE ART AND CULTURE OF THE DYNASTIC
PERIODS MENTIONED.

<u>DEVELOPMENT OF CAVES</u> AS CAN BE IMAGINED, IN THE ABSENCE OF A PLAN, THE EXCAVATIONS DURING THE DIFFERENT AGES TENDED TO BE HAPHAZARD. CAREFUL EXAMINATION, HOWEVER, HAS REVEALED THAT THE CAVES WERE STARTED AT THE CENTRE, AROUND THE KU HAN BRIDGE (古漢橋), AND DEVELOPED FIRST OF ALL TOWARDS THE SOUTH AND THE NORTH THEN UPWARDS.

RECENT INVESTIGATION HAS SET DOWN THE DYNASTIC PERIOD AND THE NUMBER OF CAVES CONSTRUCTED AS FOLLOWS:—

					NUMBER OF CAVES
WEI	(CONSTRUCTED IN	534 – 556	A.D.)		20
SUEI	("	" 581	617	")	88
TANG	("	" 618	936	")	177
FIVE DYNASTIES	("	" 907	959	")	29
SUNG	("	" 960	1276	")	102
YUAN	("	" 1277	1367	")	7
CHING	("	" 1640	1911	")	2

A THOROUGH KNOWLEDGE OF BUDDHIST LORE IS
ESSENTIAL TO LINK UP THE STORY IN ALL ITS
PHASES. WITHOUT, HOWEVER, BEING AN ANTIQUARIAN
STUDENT THE TRAVELLER CAN EASILY RECOGNISE
THE ARTISTIC AND ARCHAEOLOGICAL VALUE OF
THE FINE FRESCO PAINTINGS AND STUCCO
SCULPTURES OF THE CH'IEN FU TUNG.

J. BLACK
MARCH 1948